Studies in Rhetorics and Feminisms

Series Editors, Cheryl Glenn and Shirley Wilson Logan

Feminist Rhetorical Science Studies

Human Bodies, Posthumanist Worlds

Edited by Amanda K. Booher and Julie Jung

Southern Illinois University Press • *Carbondale*

Southern Illinois University Press
www.siupress.com

Copyright © 2018 by the Board of Trustees,
Southern Illinois University
Chapter 1 copyright © 2018 by Kyle P. Vealey and Alex Layne; chapter 3
 copyright © 2018 by Catherine C. Gouge; chapter 4 copyright © 2018 by
 Jennifer L. Bay; chapter 5 copyright © 2018 by Jordynn Jack; chapter 6
 copyright © 2018 by Daniel J. Card, Molly M. Kessler, and S. Scott Graham
All rights reserved
Printed in the United States of America

21 20 19 18   4 3 2 1

The publication of this book was made possible in part through the generous
 support of Illinois State University's English department and the College of
 Arts and Sciences.

Cover illustration: microscopic structure of bone as 3-D rendering (cropped),
 by Andrus Ciprian. Shutterstock

Cover design: Lydia Morris

Library of Congress Cataloging-in-Publication Data
Names: Booher, Amanda K., editor. | Jung, Julie, 1966– editor.
Title: Feminist rhetorical science studies : human bodies, posthumanist
 worlds / edited by Amanda K. Booher and Julie Jung.
Description: Carbondale : Southern Illinois University Press, [2018] |
 Series: Studies in rhetorics and feminisms | Includes bibliographical
 references and index.
Identifiers: LCCN 2017017874 | ISBN 9780809336333 (paperback) |
 ISBN 9780809336340 (e-book)
Subjects: LCSH: Feminism and science. | Feminism—Language. |
 Science—Language. | Feminist theory. | Communication in science. |
 BISAC: LANGUAGE ARTS & DISCIPLINES / Rhetoric. | SOCIAL SCIENCE /
 Women's Studies.
Classification: LCC Q130 .F464 2018 | DDC 500.82—dc23
LC record available at https://lccn.loc.gov/2017017874

Printed on recycled paper. ♻

Contents

Acknowledgments vii

Prologue 1

Introduction: Situating Feminist Rhetorical Science Studies 18
Amanda K. Booher and Julie Jung

1. Of Complexity and Caution: Feminism, Object-Oriented Ontology, and the Practices of Scholarly Work 50
Kyle P. Vealey and Alex Layne

2. Flat Ontologies and Everyday Feminisms: Revisiting Personhood and Fetal Ultrasound Imaging 84
Jen Talbot

3. "The Inconvenience of Meeting You": Rereading Non/Compliance, Enabling Care 114
Catherine Gouge

4. Mattering Gender: Technical Communication and Human Materiality 141
Jennifer Bay

5. How Good Brain Science Gets That Way: Reclaiming the Scientific Study of Sexed and Gendered Brains 164
Jordynn Jack

6. Representing without Representation: A Feminist New Materialist Exploration of Federal Pharmaceutical Policy 183
Daniel J. Card, Molly M. Kessler, and S. Scott Graham

7. Embodied Vernacularity at the FDA: Feminism, Epistemic Authority, and Biomedical Activism 205
Liz Barr

8. Becoming-Thinking Otherwise, Rhetorically 227
Amanda K. Booher and Julie Jung

Contributors 249

Index 251

Acknowledgments

We are indebted to series editors Cheryl Glenn and Shirley Wilson Logan as well as the organizers of the 2013 Feminisms and Rhetorics Conference (where the idea for this collection first took hold) for mentoring feminist rhetorical scholarship and modeling how to do it ethically and well. We thank Kristine Priddy, our acquisitions editor at Southern Illinois University Press, for her guidance and advocacy, and our anonymous reviewers, whose careful readings have made this book much stronger. We also thank our contributors, without whom this collection would not exist. They have listened respectfully to revision suggestions coming from multiple corners, and we are grateful for their willingness to receive critique and to revise substantively in response to it. Their sustained professionalism over these past several years has made our work infinitely easier than it might otherwise have been, and we sincerely hope the final product does their hard work justice.

I am deeply grateful to Julie for her trust, patience, and general amazingness. I'm honored to have worked on this project with such an incredible scholar, thinker, and mentor. As always, I'm thankful for the support and insights of, and opportunities to think with, Kristen Moore, Justin Hodgson, Jason Helms, Joshua Hilst, and Joshua Abboud. Finally, my eternal love and appreciation goes to my family, most of all Dennis P. Spinks, our amazing Amelia K. S. Booher (who came into being throughout the process of this book), and Linus-the-dog (whose being passed on toward the end). My heart overflows with all of you.

Amanda

I thank Amanda, whose intelligence, kindness, and all-around good vibes have made working on this project together a true pleasure. I'm also grateful for the support of a sabbatical research leave sponsored by Illinois State University's English department and the College of Arts and Sciences. I remain indebted to Angela M. Haas, Elise Verzosa Hurley, Chris Mays, Lisa L. Phillips, Kellie Sharp-Hoskins, Kirstin Hotelling Zona, J. Scott Jordan, and Christopher Breu, friends and colleagues whose ways of being-knowing in academia and beyond have had an enormous influence on this project. My never-ending gratitude goes to Rob Isaacs, Trudy Jackson, Giselle Rodriguez, and Chachi-the-poodle, for love and support.

Julie

Feminist Rhetorical Science Studies

Prologue

The version of feminist rhetorical science studies (FRSS) enacted in this collection strives to accomplish two primary goals. The first is to build alliances among scholar-teacher-activists working in feminist rhetorics, rhetorics of science, and feminist science studies (FSS). While we will situate FSS in relation to rhetoric in more detail in our introduction, for now we note that emerging scholarship in FSS relies heavily on feminist posthumanist cultural theory, also termed "feminist new materialisms." To forge cross-disciplinary alliances, then, many of our contributors draw on this same body of scholarship to develop posthumanist approaches to rhetorics of science that are explicitly feminist. In so doing, they help us achieve the book's second goal: to challenge depoliticized uptakes of posthumanism in rhetoric studies writ large. To meet these dual goals, *Feminist Rhetorical Science Studies: Human Bodies, Posthumanist Worlds* theorizes the most recent "material turn" in the intersections of feminist rhetorics and rhetorics of science.

In general, this material turn is a response to the dominance of social constructionism, which, according to Diana Coole and Samantha Frost (2010), problematically "privileges language, discourse, [and] culture," thereby failing to "give material factors their due in shaping society" (p. 3).[1] By rethinking the relationship between discourse and matter, the material turn invites more robust theorizations of materiality that have long been a concern for rhetoricians, and feminist rhetoricians in particular. Posthumanist theories and object-oriented philosophies, for example, resonate with existing scholarship that studies rhetoric in excess of human symbolic action. Such inquiries consider how, for instance, embodiment is rhetorical; rhetoric is embodied; embodiment exceeds normative narratives and thus resists discursive incorporation; embodiment functions as a prediscursive site of rhetorical agency; and rhetorical interactions can be theorized as dynamic networks involving material and discursive objects.[2]

The significance of the material can also be seen in the return to questions of methodology and the concomitant need to understand research as embodied practice situated in space and time (Royster & Kirsch, 2012; Schell & Rawson, 2010). Feminist archival research, in particular, emphasizes the affordances of conceptualizing rhetorical historical inquiry as the study of interactions between discourses and material artifacts.[3] Paramount in all these inquiries is the impetus to theorize rhetoric as being inextricably connected to issues of language and human agency, but not limited to only them.

On the other hand, while emerging theories of materiality are attuned to important issues in contemporary rhetoric scholarship, they also raise concerns for scholars doing work in feminist rhetorics. Specifically, as several of our contributors argue, much scholarship in posthumanism and object-oriented ontology/object-oriented rhetoric posits that human and nonhuman beings (animals, plants, objects) operate on equal immanent planes, affecting and being affected by each other in (potentially) equal measure. The purpose of such an approach is twofold: (*1*) to decenter the human and particularly Enlightenment notions of the human as dominant over nature and objects; and (*2*) to see and understand the agency of nonhuman objects as significant in their own right in order to better understand the relationships of humans and nonhumans in the world.

While such an approach offers important challenges to an ideology of human exceptionalism, we nevertheless see some limitations as they relate to the "feminist" in FRSS. Specifically (and as explicated more fully in the first three chapters of this collection), mainstream posthumanism's emphasis on the symmetry between humans and nonhumans and object-oriented philosophy's claim that everything is an object eclipse issues of difference within the category of "human" and thus elide questions about how such distinctions get made, why they get made, and to what effect. Too often, posthumanist and object-oriented studies fail to ask questions of race, culture, gender, disability, or other features that differentiate humans from each other (and thus from their relationships with nonhumans) in important, impactful ways.

We are also concerned about the ethics that potentially get lost when these features of complex humanity are not addressed. As historically underrepresented, underappreciated, and underprotected groups, those humans who are *not* white, male, able-bodied, straight, and so on are also often underhumanized, not easily given the political, social, cultural, and other rights and protections afforded other humans. It seems to us that some scholars of the posthuman/object-oriented are quick to throw off the bodily compartments that orient them in the material world. As such, some posthumanist and object-oriented approaches risk reinscribing a privileged position that allows one to minimize the body—its agencies, variances, e/affects—by placing it on an immanent plane with all other objects. To contest this privilege, this book attends to the rhetorical complexities of differentiated embodiment (in terms of both its production and effects) by forging a collective via the heading "feminist rhetorical science studies."

As a book title, *Feminist Rhetorical Science Studies* requires significant unpacking, for its intellectual genealogy is messy, layered, complex—and complexity demands selection. As systems theorist Niklas Luhmann (1996) puts it, "complexity means being forced to select; being forced to select means contingency; and contingency means risk" (p. 25). Contemplating alternatives is a useful strategy not only for imagining how things might be conceived differently but also for testing commitments to selections already made. For example, we could have titled this collection *Feminisms, Rhetorics, Sciences*, which signals plurality and the existence of differences within each category in a way that our selected title does not. And with only commas separating the categories, there are spaces out of which different sorts of distinctions and connections can be made. Yet a major goal of this book is to put feminist rhetorical theory in conversation with the interdisciplinary field of FSS. By adopting FSS's adjectival use of *feminist*, our title works rhetorically to establish a readily apparent theoretical linkage between the two fields. It also signals a shared commitment to extending feminist inquiry beyond a woman-only focus in order to combat gender essentialism and examine intersections among different forms of complex embodiment.[4] Additionally, using both *feminist* and *rhetorical* adjectivally situates "science studies" as their shared object of inquiry.

However, our addition of *rhetorical* to FSS is not without risk. As a "heterogeneous and amorphous body of work" with "porous interdisciplinary boundaries" (Subramaniam, 2009, p. 953), FSS is committed to "tackling the last and most difficult barrier to interdisciplinarity—between the humanities and the natural/physical sciences" (Mayberry, Subramaniam, & Weasel, 2001, p. 2). As such, our title's singling out of rhetoric could construct a categorical boundary that may seem at odds with FSS's goal of boundary transgression, if not dissolution.[5] For this reason, we employ the adjectival form, *rhetorical*, to emphasize how feminist rhetorical theory, like feminist theory, offers theoretical frameworks and methodological approaches capable of traveling across and contributing to scholarship in diverse disciplines.

Our intention in using *Feminist Rhetorical Science Studies* is thus neither to claim a distinctly "new" field nor to suggest that "science" exists as a monolithic singularity. Indeed, scholarship in the Rhetoric of Science, Technology, and Medicine (RSTM) evidences rhetoricians' resistance to conceptualizing "science" as a discrete object of inquiry. Instead, our intention is an explicitly rhetorical one: to develop one title for feminist scholarship in the rhetoric of science that is capable of responding to the exigencies that motivate our project while also enacting innovative connections across multiple fields.[6] Such a goal is in keeping with the kind of critical connection-making that characterizes scholarship in RSTM. As Jeanne Fahnestock (2013) notes, research in RSTM is related not only to "the big three—history, sociology, and philosophy" but also to "linguistics, literary studies, visual studies and art history, journalism, public relations, as well as applied fields like Public Health or Environmental Studies." Even this is merely a short list.

To complicate matters further, each object of rhetorical inquiry identified in the acronym RSTM—science, technology, and medicine—is itself a heterogeneous field that sustains important interdisciplinary connections as it simultaneously forges its own identity. Scholarship in the rhetoric of technology, or rhetorics *and* technologies, for example, manifests complicated intersections of two nebulous but vital terms. In "The Rhetoric of Technology as a Rhetorical Technology," John A. Lynch and William J. Kinsella (2013) reflect on these complications via fundamental underlying issues of definition (in other words, asking "what is technology?" reveals answers as vast as asking "what is rhetoric?"), noting how the term *technology* can variously "refer to discrete artifacts, the systems and practices surrounding the use of said artifacts, and the application of scientific principles to practice (i.e., engineering)." Thus, work within this field covers widely different questions and approaches. For example, research in rhetoric of/and technology studies user practices associated with different technologies (especially in, but not limited to, the classroom); the histories (chronologically, socially, culturally) of technologies; humans' relationships to/with technologies; and debates about the very essence of technology (and humanity) itself (and whether *essence* is even an appropriate term).[7]

Not surprisingly, then, questions of identity emerge for rhetoricians working in RSTM. According to Fahnestock (2013), RSTM

> is (1) distinctive for its attention to the choices that rhetors make against a background of possibilities, both in malleably defined contexts as well as at every possible level of formal analysis; and (2) different in pursuing the "why behind" and "what follows from" these choices.

Additional points of distinction include questioning for whom we are writing— those audiences primarily within or beyond the field of rhetorics (Ceccarelli,

2013); the genres that are relevant to the field (and how *genre* itself is defined) (Miller & Fahnestock, 2013); the lack of international focus in RSTM in public discourse (Condit, 2013); and the grounding relation to "invention" in rhetoric and technology (Lynch & Kinsella, 2013). Given the "new institutional environment[s]" associated with changing structures in the university and the funding of research, scholars in RSTM are also concerned with how they can substantively participate "in interdisciplinary research projects in which rhetoric functions as a significant contributor to research, outreach, and policy formation" (Herndl & Cutlip, 2013). What is not particularly highlighted here, and what we hope to bring forth in this collection, is a specific focus on *feminism* and RSTM, and, relatedly, to areas of scholarship that query gender/sex, sexuality, race, culture, disability, and other aspects of embodiment vital to the various topics in FSS.

By establishing an alliance between FSS and rhetorical inquiry, this book also aims to intervene in problematic uptakes of posthumanism in the field of rhetoric studies broadly conceived. Specifically, the inescapable politics of mattering distinguishes *Feminist Rhetorical Science Studies* from posthumanist rhetoric scholarship that theorizes "embodiment and agency . . . as depoliticized effects of complexity or networked realities" (Micciche, 2014, p. 491). With this problem in mind, our book is committed to attending to that which feminist critics contend is too easily ignored: those practices that sustain asymmetrical relations of power between differentially embodied human beings.

On Reading *Feminist Rhetorical Science Studies*

Contributors Kyle P. Vealey and Alex Layne propose a methodology of rhetorical reverberations, which is primarily concerned with ethical writing and citation practices, yet we believe it can motivate ethical reading practices as well. As such, we invite our readers to enact the contributors' methodology (outlined in chapter 1) and help build new ontologies for feminist rhetorics and FSS by (*1*) paying attention to the citations that do and do not appear in this collection; (*2*) tracking the most glaring exclusions; and (*3*) producing scholarship that makes those exclusions matter.

This scholarly practice is, of course, one with which feminists of color are well familiar, and its potential affordances for posthumanist rhetorics are compellingly exemplified in Gabriela Raquel Ríos's "Cultivating Land-Based Literacies and Rhetorics" (2015). Noting how contemporary scholarship in object-oriented rhetoric challenges "the dichotomous philosophy that would separate environment/human/mind," Ríos observes that "Indigenous philosophies and rhetorics have always resisted such dichotomies" (p. 65). By pointing out how most scholars in posthumanist rhetorics—feminist and not—have ignored non-Western philosophies, and, further, by demonstrating how those very philosophies can powerfully contribute to rhetorical theories and practices, Ríos makes them matter differently. Vealey and Layne suggest—and we agree—that *every* scholar shares the ethical obligation of engaging in the kind of scholarship Ríos models, which, in making previously excluded knowledges matter, helps to remake the field. Among other things, such engagement can help make apparent that while interdependence is indeed an ontological fact of human existence, some relational configurations have come to matter more than others. We believe that understanding how this differential mattering happens, for whom, and why is the work of rhetoric.

Modeling this approach, we briefly remark here on two important exclusions that appear in this book. Our goal of foregrounding the complexities of differentially valued human embodiment led to the development of feminist posthumanist frameworks that contend with how different kinds of *human* bodies come to matter differently. While we believe such a focus provides a much-needed corrective to posthumanist rhetorics that depoliticize human mattering, we also recognize what this focus ignores: none of our chapters develop a framework that makes explicit connections between scholarship in RSTM and that in critical animal studies. Although rhetoric scholars have begun to challenge the anthropocentrism of Western philosophies and rhetorics by bringing the mattering of nonhuman animals to the fore (see, for example, Hawhee, 2011; Ridolfo, 2015; Walters, 2010, 2013, 2014; Worsham, 2015), clearly much more of this kind of work needs to be done in relation to rhetorics of science specifically. We hope our readers will continue the conversation taken up in this collection, extending or complicating one or more of the approaches developed here in order to respond to questions such as: How can feminist posthumanist rhetorical frameworks theorize medical testing on nonhuman animals as specific kinds of meaning-matter entanglements? How do human researchers, nonhuman animals, laboratory apparatuses, and ideologies, such as speciesism, interact in ways that render the nonhuman animal a particular type of research "subject"? What might the implications of a rethinking of this subjectivity be with respect to (to take just one example) the ways in which institutional review boards determine who belongs to a "vulnerable population" and why?

A second notable exclusion is the absence of contributions that develop feminist posthumanist rhetorical theories that challenge posthumanism's Western-centrism by drawing on scholarship in Indigenous and postcolonial studies. As Ríos's important article makes clear, the fact that human and nonhuman animals, objects, and lands are enmeshed is not a recent Western "discovery." Yet to fail to recognize non-Western intellectual traditions is to engage in what philosopher José Medina (2013) terms "epistemic neglect," the capacity to ignore alternative ways of knowing for purposes of sustaining privilege and existing relations of power (p. 34). Together, Ríos's and Medina's arguments provide a potent reminder that despite posthumanism's investment in theorizing ontology, *feminist* posthumanist scholars must also remain committed to interrogating issues of epistemology—who is participating in making knowledge that counts, for whom, how, and why?

We hope our readers will keep these questions in mind as they read across the chapters that follow. As we engaged in the work of compiling and editing this collection, we asked these questions of ourselves, which in turn prompted us to notice one significant exclusion that appears in our own introduction: cyborg feminism's indebtedness to scholarship by feminists of color and of the global south. In "New Sciences: Cyborg Feminism and the Methodology of the Oppressed," Chela Sandoval (2000) ponders the tendency among academic feminists to elide the ways in which Donna Haraway's cyborg feminism emerges from

> techniques and terminologies of US Third World cultural forms, from Native American concepts of "trickster" and "coyote" being, to "mestiza" or the category of "women of color," until the body of the feminist cyborg becomes clearly articulated with the material and psychic positionings of US Third World feminism. (p. 378)

It would be tempting, of course, to explain away this elision by blaming Haraway for it, but Sandoval doesn't let her readers off so easily. Pointing out that "Haraway has been very clear about these intellectual lineages and alliances" (p. 377), and, too, that Haraway has acknowledged how "her work inadvertently contributed to th[e] tendency to elide the specific theoretical contributions of US Third World feminist criticism," Sandoval asks: "Why has feminist theory been unable to recognize US Third World feminist criticism as a mode of cultural theory?" (p. 381).[8] The implications of Sandoval's question become more salient when she explains how "colonized peoples of the Americas have already developed the cyborg skills required for survival under techno-human conditions" (p. 375) and thus have already enacted the nonessentializing, insurgent subject theorized in Haraway's cyborg.

In acknowledging this history and drawing on it to theorize a methodology of the oppressed, Sandoval does not advocate that we abandon cyborg feminism or devalue Haraway's significant contributions to feminist theory. Rather, she calls on us to ask: how and why has Haraway's cyborg come to matter so much to so many, while Sandoval's methodology of the oppressed has not? By making this elision apparent, Sandoval also points to a possible tactic for reconstituting what matters in FRSS: whenever we find ourselves as scholars oriented toward Haraway's cyborg feminism, we might turn toward Sandoval's methodology of the oppressed instead.

This, then, is how we conclude, for now: with a reminder that what the world is now is in the process of becoming something else. Yet this inexorable becoming does not render it impervious to intervention. Our work as scholars and teachers—what our work is, what it does, and how we do it—has a hand in that becoming. The selections we make today help bring into being some worlds to the exclusion of others. Our hopes for the selections we've made here, for the work of this collection, are many. Mostly, though, we hope to help bring into being a collective of feminist scholars of science who are alive to the possibility of being-knowing otherwise.

Notes

1. A note on our rhetorical use of verb tense: we follow APA citation guidelines to build bridges with scholars in the sciences and social sciences; when situating existing research within our argument, however, we ignore APA guidelines that call for authors to use past tense. We do so in order to foreground our posthumanist orientation that meanings of texts (and research findings) *become* via material-discursive intra-actions.

2. For a discussion of how embodiment is rhetorical, see Lay, Gurak, Gravon, and Myntti, 2000; Selzer and Crowley, 1999; Wilson and Lewiecki-Wilson, 2001. For how rhetoric is embodied, see Dingo, 2012; Dolmage, 2014; Hawhee, 2009; McKerrow, 1998; Pezzullo, 2007. For how embodiment exceeds normative narratives and thus resists discursive incorporation, see Hesford, 2011; Wiederhold, 2002; Worsham, 1998. For how embodiment functions as a prediscursive site of rhetorical agency, see Cooper, 2011; Davis, 2010. And for how rhetorical interactions can be theorized as dynamic networks involving material and discursive objects, see Edbauer, 2005; Hawk, 2007; Propen, 2012; Rice, 2012; Rickert, 2013.

3. See Anderson and Enoch, 2013; Enoch, 2008, 2011; Jack, 2009; Kirsch and Rohan, 2008; Ramsey, Sharer, L'Eplattenier, and Mastrangelo, 2009; Wells, 2010.

4. Sarah S. Richardson (2010) explains in her historical overview of feminist philosophy of science that "the questions and frameworks of feminist scholarship . . . extend beyond issues of gender to analysis of ideology in scientific work on human difference in areas such as race, ethnicity, sexuality, and disability"

(p. 358). Cho, Crenshaw, and McCall (2013) argue that this type of approach is characterized by

> an intersectional way of thinking about the problem of sameness and difference and its relation to power. This framing—conceiving of categories not as distinct but as always permeated by other categories, fluid and changing, always in the process of creating and being created by dynamics of power—emphasizes what intersectionality does rather than what intersectionality is. (p. 795)

The adoption of intersectional methodologies in feminist rhetorics is well established. See, for example, Calafell, 2014; Chávez and Griffin, 2012; Glenn and Lunsford, 2014; Royster and Kirsch, 2012; Sano-Franchini, Sackey, and Pigg, 2011; Schell and Rawson, 2010.

5. Singling out rhetoric as a separate field might also have the added benefit of exposing more scholars in science studies to scholarship in rhetoric, a need identified by rhetoricians of science. In her contribution to a special issue of *Poroi* dedicated to charting new directions for scholarship in the Rhetoric of Science, Technology, and Medicine (RSTM), Jeanne Fahnestock (2013) observes that "scholars in related fields do not routinely access or acknowledge rhetorical studies." Given that "some colleagues in other fields still see rhetoric in negative terms," issue editor Lisa Keränen (2013) suggests that our "low interdisciplinary influence" could in part be due to "a misunderstanding of our key term." It's also worth remembering that FSS is a field of inquiry greatly influenced by feminist philosophy of science. As such, many scholars doing theoretical work in FSS likely have not been trained in rhetoric and thus may be unable to see how knowledge of it can productively contribute to their research. While our title won't work any miracles, it may help increase rhetoric's visibility among scholars conducting searches using the terms *feminist science* and *science studies*.

6. Although a comprehensive review of the literature in RSTM is beyond the scope of this prologue, we do offer a very partial listing of some areas of inquiry and examples of feminist scholarship within each that we think are especially relevant to doing work in FRSS as we have defined it. In histories of rhetoric, see Hallenbeck, 2012; Hayden, 2013; Jack, 2009; Skinner, 2014. In disability rhetorics, see Booher, 2010; Jack, 2011, 2012; Moeller, 2014; Owens, 2009. In technical communication, see Frost, 2014; Haas, 2012; Lay et al., 2000; Tillery, 2005. In digital and visual rhetorics, see Blair, Gajjala, and Tulley, 2008; Enoch and Bessette, 2013; Gries, 2015; Haas, 2007; Propen, 2012; Rhodes, 2004. In the rhetoric of medicine, see Brasseur, 2012; Dubriwny and Ramadurai, 2013; Happe, 2013, "Body of Race"; Hausman, 2013; Wells, 2001, 2010; Willard, 2005;

Woods, 2013. In transnational feminist rhetorics, see Dingo, 2012; Hesford, 2011; Queen, 2008; Wang, 2013. In environmental rhetorics, see Pezzullo, 2007; Ríos, 2015. Scholarship in feminist rhetorical scholarship in the rhetoric of science also examines a diverse range of bodies, objects, and phenomena. For research on pregnancy, childbirth, breastfeeding, and motherhood, see Dubriwny and Ramadurai, 2013; Hausman, 2013; Hurt, 2011; Koerber, 2006; Lay, 2000; Owens, 2009; Spoel, 2007; Thornton, 2014. For research on scientific racism and racial ideology, see Happe, 2013, *Material Gene*; Richardson, 2014. For research on illness and disease, see Duerringer, 2013. For studies of scientific writing and discourse, see Segal, 1995; Tillery, 2005; Wells, 2001. For research on abortion and reproductive rights, see Gibson, 2008.

7. Again, though a comprehensive review of the literature in technology and/or rhetorics is beyond the scope of this work, the following is a brief listing of relevant works. For studying the practice of using different technologies, see Arola, Sheppard, and Ball, 2014; Bernhardt, 2013; Brooke, 2009; Haas, 2012; Lynch, 2012. For histories of technologies, see Ceccarelli, 2013; Durack, 1997; Fahnestock, 2013; Hallenbeck, 2012; Herndl and Cutlip, 2013; Katz, 1992; Kimball, 2006; Lay et al., 2000; Lynch and Kinsella, 2013. For humans' relationship to/with technologies, see Barnett, 2010; Dobrin, 2015; Jung, 2014; Mol, 2002; Morton, 2011; Muckelbauer and Hawhee, 2000; Rickert, 2013; Spinuzzi, 2012; Zemlicka, 2013.

8. For critiques of Haraway's appropriation of scholarship by feminists of color, see Moya, 1997, pp. 128–133, and Robson, 1998, p. 54.

References

Anderson, D., & Enoch, J. (Eds.). (2013). *Burke in the archives: Using the past to transform the future of Burkean studies.* Columbia: University of South Carolina Press.

Arola, K. L., Sheppard, J., & Ball, C. E. (2014). *Writer/designer: A guide to making multimodal projects.* New York, NY: Bedford/St. Martin's.

Barnett, S. (2010). Toward an object-oriented rhetoric [Review of the books *Tool-being: Heidegger and the metaphysics of objects* and *Guerrilla metaphysics: Phenomenology and the carpentry of things*, by Graham Harman]. *Enculturation, 7.* Accessed 30 June 2017. Retrieved from http://enculturation.net/toward-an-object-oriented-rhetoric

Bernhardt, S. A. (2013). Review essay: Rhetorical technologies, technological rhetorics. *College Composition and Communication, 64*(4), 704–720.

Blair, K., Gajjala, R., & Tulley, C. (Eds.). (2008). *Webbing cyberfeminist practice: Communities, pedagogies, and social action.* New York, NY: Hampton Press.

Booher, A. K. (2010). Docile bodies, supercrips, and the plays of prosthetics. *IJFAB, 3*(2), 63–89.

Brasseur, L. (2012). Sonographers' complex communication during the obstetric sonogram exam: An interview study. *Journal of Technical Writing and Communication, 42*(1), 3–19.

Brooke, C. G. (2009). *Lingua fracta: Towards a rhetoric of new media.* New York, NY: Hampton Press.

Calafell, B. M. (2014). The future of feminist scholarship: Beyond the politics of inclusion. *Women's Studies in Communication, 37*(3), 266–270.

Ceccarelli, L. (2013). To whom do we speak? The audiences for scholarship on the rhetoric of science and technology. *Poroi, 9*(1). Accessed 30 June 2017. Retrieved from http://ir.uiowa.edu/cgi/viewcontent.cgi?article=1151&context=poroi

Chávez, K. R., & Griffin, C. L. (Eds.). (2012). *Standing in the intersection: Feminist voices, feminist practices in communication studies.* Albany: State University of New York Press.

Cho, S., Crenshaw, K. W., & McCall, L. (2013). Toward a field of intersectionality studies: Theory, applications, and praxis. *Signs, 38*(4), 785–810.

Condit, C. M. (2013). "Mind the gaps": Hidden purposes and missing internationalism in scholarship on the rhetoric of science and technology in public discourse. *Poroi, 9*(1). Accessed 30 June 2017. Retrieved from http://ir.uiowa.edu/poroi/vol9/iss1/3/

Coole, D., & Frost, S. (Eds.). (2010). *New materialisms: Ontology, agency, and politics.* Durham, NC: Duke University Press.

Cooper, M. M. (2011). Rhetorical agency as emergent and enacted. *College Composition and Communication, 62*(3), 420–449.

Davis, D. (2010). *Inessential solidarity: Rhetoric and foreigner relations.* Pittsburgh, PA: University of Pittsburgh Press.

Dingo, R. (2012). *Networking arguments: Rhetoric, transnational feminism, and public policy writing.* Pittsburgh, PA: University of Pittsburgh Press.

Dobrin, S. I. (Ed.). (2015). *Writing posthumanism, posthuman writing.* Anderson, SC: Parlor Press.

Dolmage, J. T. (2014). *Disability rhetoric.* Syracuse, NY: Syracuse University Press.

Dubriwny, T. N., & Ramadurai, V. (2013). Framing birth: Postfeminism in the delivery room. *Women's Studies in Communication, 36*(3), 243–266.

Duerringer, C. M. (2013). Winking and giggling at creeping death: Thanatophobia and the rhetoric of Save the Ta-Tas. *Journal of Communication Inquiry, 37*(4), 344–363.

Durack, K. T. (1997). Gender, technology, and the history of technical communication. *Technical Communication Quarterly, 6*(3), 249–260.

Edbauer, J. (2005). Unframing models of public distribution: From rhetorical situation to rhetorical ecologies. *Rhetoric Society Quarterly, 35*(4), 5–24.

Enoch, J. (2008). *Refiguring rhetorical education: Women teaching African American, Native American, and Chicano/a students, 1865–1911.* Carbondale: Southern Illinois University Press.

Enoch, J. (2011). Finding new spaces for rhetorical research. *Rhetoric Review, 30*(2), 115–117, 131.

Enoch, J., & Bessette, J. (2013). Meaningful engagements: Feminist historiography and the digital humanities. *College Composition and Communication, 64*(4), 634–660.

Fahnestock, J. (2013). Promoting the discipline: Rhetorical studies of science, technology, and medicine. *Poroi, 9*(1). Accessed 30 June 2017. Retrieved from http://ir.uiowa.edu/poroi/vol9/iss1/6/

Frost, E. (2014). An apparent feminist approach to transnational technical rhetorics: The ongoing work of Nujood Ali. *Peitho, 16*(2), 183–199.

Gibson, K. L. (2008). The rhetoric of *Roe v. Wade*: When the (male) doctor knows best. *Southern Communication Journal, 73*(4), 312–331.

Glenn, C., & Lunsford, A. A. (Eds.). (2014). *Landmark essays on rhetoric and feminism, 1973–2000.* New York, NY: Routledge.

Gries, L. E. (2015). *Still life with rhetoric: A new materialist approach for visual rhetorics.* Boulder, CO: Utah State University Press.

Haas, A. M. (2007). Wampum as hypertext: An American Indian intellectual tradition of multimedia theory and practice. *Studies in American Indian Literatures, 19*(4), 77–100.

Haas, A. M. (2012). Race, rhetoric, and technology: A case study of decolonial technical communication theory, methodology, and pedagogy. *Journal of Business and Technical Communication, 26*(3), 277–310.

Hallenbeck, S. (2012). Toward a posthuman perspective: Feminist rhetorical methodologies and everyday practices. *Advances in the History of Rhetoric, 15*(1), 9–27.

Happe, K. E. (2013). The body of race: Toward a rhetorical understanding of racial ideology. *Quarterly Journal of Speech, 99*(2), 131–155.

Happe, K. E. (2013). *The material gene: Gender, race, and heredity after the Human Genome Project.* New York: New York University Press.

Hausman, B. L. (2013). Breastfeeding, rhetoric, and the politics of feminism. *Journal of Women, Politics, and Policy, 34*(4), 330–344.

Hawhee, D. (2009). *Moving bodies: Kenneth Burke at the edges of language.* Columbia: University of South Carolina Press.

Hawhee, D. (2011). Toward a bestial rhetoric. *Philosophy and Rhetoric, 44*(1), 81–87.

Hawk, B. (2007). *A counter-history of composition: Toward methodologies of complexity.* Pittsburgh, PA: University of Pittsburgh Press.

Hayden, W. (2013). *Evolutionary rhetoric: Sex, science, and free love in nineteenth-century feminism.* Carbondale: Southern Illinois University Press.

Herndl, C. G., & Cutlip, L. L. (2013). "How can we act?" A praxiographical program for the rhetoric of technology, science, and medicine. *Poroi, 9*(1). Accessed 30 June 2017. Retrieved from http://ir.uiowa.edu/cgi/viewcontent.cgi?article=1163&context=poroi

Hesford, W. S. (2011). *Spectacular rhetorics: Human rights visions, recognitions, feminisms.* Durham, NC: Duke University Press.

Hurt, N. E. (2011). Legitimizing "baby brain": Tracing a rhetoric of significance through science and the mass media. *Communication and Critical/Cultural Studies, 8*(4), 376–398.

Jack, J. (2009). *Science on the home front: American women scientists in World War II.* Urbana: University of Illinois Press.

Jack, J. (2011). "The extreme male brain?" Incrementum and the rhetorical gendering of autism. *Disability Studies Quarterly, 31*(3). Accessed 30 June 2017. Retrieved from http://dsq-sds.org/article/view/1672/1599

Jack, J. (2012). Gender copia: Feminist rhetorical perspectives on an autistic concept of sex/gender. *Women's Studies in Communication, 35*(1), 1–17.

Jung, J. (2014). Systems rhetoric: A dynamic coupling of explanation and description. *Enculturation, 17*(1). Accessed 30 June 2017. Retrieved from http://www.enculturation.net/systems-rhetoric

Katz, S. B. (1992). The ethic of expediency: Classical rhetoric, technology, and the Holocaust. *College English, 54*(3), 255–275.

Keränen, L. (2013). Conspectus: Inventing futures for the rhetoric of science, technology, and medicine. *Poroi, 9*(1). Accessed 30 June 2017. Retrieved from http://ir.uiowa.edu/poroi/vol9/iss1/1/

Kimball, M. A. (2006). Cars, culture, and tactical technical communication. *Technical Communication Quarterly, 15*(1), 67–86.

Kirsch, G. E., & Rohan, L. (Eds.). (2008). *Beyond the archives: Research as a lived process.* Carbondale: Southern Illinois University Press.

Koerber, A. (2006). Rhetorical agency, resistance, and the disciplinary rhetorics of breastfeeding. *Technical Communication Quarterly, 15*(1), 87–101.

Lay, M. M. (2000). *The rhetoric of midwifery: Gender, knowledge, and power.* New Brunswick, NJ: Rutgers University Press.

Lay, M. M., Gurak, L. J., Gravon, C., & Myntti, C. (Eds.). (2000). *Body talk: Rhetoric, technology, reproduction.* Madison: University of Wisconsin Press.

Luhmann, N. (1996). *Social systems.* (J. Bednarz Jr., Trans.). Palo Alto, CA: Stanford University Press.

Lynch, J. A., & Kinsella, W. J. (2013). The rhetoric of technology as a rhetorical technology. *Poroi, 9*(1). Accessed 30 June 2017. Retrieved from http://ir.uiowa.edu/poroi/vol9/iss1/13/

Lynch, P. (2012). Composition's new thing: Bruno Latour and the apocalyptic turn. *College English, 74*(5), 458–476.

Mayberry, M., Subramaniam, B., & Weasel, L. H. (Eds.). (2001). *Feminist science studies: A new generation.* New York: Routledge.

McKerrow, R. E. (1998). Corporeality and cultural rhetoric: A site for rhetoric's future. *Southern Communication Journal, 63*(4), 315–328.

Medina, J. (2013). *The epistemology of resistance: Gender and racial oppression, epistemic injustice, and resistant imaginations.* Oxford, UK: Oxford University Press.

Micciche, L. R. (2014). Writing material. *College English, 74*(6), 488–505.

Miller, C. R., & Fahnestock, J. (2013). Genres in scientific and technical rhetoric. *Poroi, 9*(1). Accessed 30 July 2017. Retrieved from http://ir.uiowa.edu/poroi/vol9/iss1/12/

Moeller, M. (2014). Pushing boundaries of normalcy: Employing critical disability studies in analyzing medical advocacy websites. *Communication Design Quarterly, 2*(4), 52–80.

Mol, A. (2002). *The body multiple: Ontology in medical practice.* Durham, NC: Duke University Press.

Morton, T. (2011). Sublime objects. In M. Austin, P. J. Ennis, F. Gironi, & T. Gokey (Eds.), *Speculations II* (pp. 207–227). Brooklyn, NY: Punctum Books.

Moya, P. M. L. (1997). Postmodernism, "realism," and the politics of identity: Cherríe Moraga and Chicana feminism. In M. J. Alexander & C. T. Mohanty (Eds.), *Feminist genealogies, colonial legacies, democratic futures* (pp. 125–150). New York, NY: Routledge.

Muckelbauer, J., & Hawhee, D. (2000). Posthuman rhetorics: "It's the future, Pikul." *JAC, 20*(4), 767–774.

Owens, K. H. (2009). Confronting rhetorical disability: A critical analysis of women's birth plans. *Written Communication, 26*(3), 247–272.

Pezzullo, P. C. (2007). *Toxic tourism: Rhetorics of pollution, travel, and environmental justice.* Tuscaloosa: University of Alabama Press.

Propen, A. D. (2012). *Locating visual-material rhetorics: The map, the mill, and the GPS.* Anderson, SC: Parlor Press.

Queen, M. (2008). Transnational feminist rhetorics in a digital world. *College English, 70*(5), 471–489.

Ramsey, A. E., Sharer, W. B., L'Eplattenier, B., & Mastrangelo, L. S. (Eds.). (2009). *Working in the archives: Practical research methods for rhetoric and composition.* Carbondale: Southern Illinois University Press.

Rhodes, J. (2004). *Radical feminism, writing, and critical agency: From manifesto to modem.* Albany: State University of New York Press.

Rice, J. (2012). *Distant publics: Development rhetoric and the subject of crisis.* Pittsburgh, PA: University of Pittsburgh Press.

Richardson, F. (2014). The eugenics agenda: Deliberative rhetoric and therapeutic discourse of hate. In M. F. Williams & O. Pimentel (Eds.), *Communicating race, ethnicity, and identity in technical communication* (pp. 7–22). Amityville, NY: Baywood.

Richardson, S. S. (2010). Feminist philosophy of science: History, contributions, and challenges. *Synthese, 177*(3), 337–362.

Rickert, T. (2013). *Ambient rhetoric: The attunements of rhetorical being.* Pittsburgh, PA: University of Pittsburgh Press.

Ridolfo, J. (2015). Rethinking human and non-human actors as a strategy for rhetorical delivery. In S. I. Dobrin (Ed.), *Writing posthumanism, posthuman writing* (pp. 174–191). Anderson, SC: Parlor Press.

Ríos, G. R. (2015). Cultivating land-based literacies and rhetorics. *LiCS, 3*(1), 60–70.

Robson, R. (1998). *Sappho goes to law school.* New York, NY: Columbia University Press.

Royster, J. J., & Kirsch, G. E. (2012). *Feminist rhetorical practices: New horizons for rhetoric, composition, and literacy studies.* Carbondale: Southern Illinois University Press.

Sandoval, C. (2000). New sciences: Cyborg feminism and the methodology of the oppressed. In D. Bell & B. M. Kennedy (Eds.), *The cybercultures reader* (pp. 374–387). London, UK: Routledge.

Sano-Franchini, J., Sackey, D., & Pigg, S. (2011). Methodological dwellings: A search for feminisms in rhetoric and composition. *Present Tense, 1*(2). Accessed 30 June 2017. Retrieved from http://www.presenttensejournal.org/v011/methodological-dwellings-a-search-for-feminisms-in-rhetoric-composition

Schell, E. E., & Rawson, K. J. (Eds.). (2010). *Rhetorica in motion: Feminist rhetorical methods and methodologies.* Pittsburgh, PA: University of Pittsburgh Press.

Segal, J. Z. (1995). Is there a feminist rhetoric of medicine? Toward a critical pedagogy for scientific writing. *Textual Studies in Canada, 7,* 109–116.

Selzer, J., & Crowley, S. (Eds.). (1999). *Rhetorical bodies.* Madison: University of Wisconsin Press.

Skinner, C. (2014). *Women physicians and professional ethos in nineteenth-century America.* Carbondale: Southern Illinois University Press.

Spinuzzi, C. (2012). Working alone, together: Coworking as emergent collaborative activity. *Journal of Business and Technical Communication, 26*(4), 399–441.

Spoel, P. (2007). A feminist rhetorical perspective on informed choice in midwifery. *Rhetor, 2.* Accessed 30 June 2017. Retrieved from http://www.cssr-scer.ca/wp-content/uploads/2015/05/Rhetor2_Spoel.pdf

Subramaniam, B. (2009). Moored metamorphoses: A retrospective essay on feminist science studies. *Signs, 34*(4), 951–980.

Thornton, D. (2014). Transformations of the ideal mother: The story of mommy economicus and her amazing brain. *Women's Studies in Communication, 37*(3), 271–291.

Tillery, D. (2005). The plain style in the seventeenth century: Gender and the history of scientific discourse. *Journal of Technical Writing and Communication, 35*(3), 273–289.

Walters, S. (2010). Animal Athena: The interspecies *metis* of women writers with autism. *JAC, 30*(3/4), 683–711.

Walters, S. (2013). Seeking, playful translators: Emotion, difference, and connection in human-animal relations. *JAC, 33*(1/2), 336–351.

Walters, S. (2014). Unruly rhetorics: Disability, animality, and new kinship compositions. *PMLA, 129*(3), 471–477.

Wang, B. (2013). Comparative rhetoric, postcolonial studies, and transnational feminisms: A geopolitical approach. *Rhetoric Society Quarterly, 43*(3), 226–242.

Wells, S. (2001). *Out of the dead house: Nineteenth-century women physicians and the writing of medicine.* Madison: University of Wisconsin Press.

Wells, S. (2010). *"Our bodies, ourselves" and the work of writing.* Stanford, CA: Stanford University Press.

Wiederhold, E. (2002). The face of mourning: Deploying grief to construct a nation. *JAC, 22*(4), 847–889.

Willard, B. E. (2005). Feminist interventions in biomedical discourse: An analysis of the rhetoric of integrative medicine. *Women's Studies in Communication, 28*(1), 115–148.

Wilson, J. C., & Lewiecki-Wilson, C. (Eds.). (2001). *Embodied rhetorics: Disability in language and culture.* Carbondale: Southern Illinois University Press.

Woods, C. S. (2013). Repunctuated feminism: Marketing menstrual suppression through the rhetoric of choice. *Women's Studies in Communication, 36*(3), 267–287.

Worsham, L. (1998). Going postal: Pedagogic violence and the schooling of emotion. *JAC, 18*(2), 213–245.

Worsham, L. (2015). Moving beyond the logic of sacrifice: Animal studies, trauma studies, and the path to posthumanism. In S. I. Dobrin (Ed.), *Writing posthumanism, posthuman writing* (pp. 19–55). Anderson, SC: Parlor Press.

Zemlicka, K. (2013). The rhetoric of enhancing the human: Examining the tropes of "the human" and "dignity" in contemporary bioethical debates over enhancement technologies. *Philosophy and Rhetoric, 46*(3), 257–279.

Introduction

Situating Feminist Rhetorical Science Studies

Amanda K. Booher and Julie Jung

*F*eminist Rhetorical Science Studies strives to develop theoretical and methodological frameworks for engaging posthumanism as feminist rhetoricians of science and scholars whose work contributes to ongoing conversations in feminist science studies (FSS). As such, this collection enacts one contingently bounded and thus necessarily incomplete version of feminist rhetorical science studies (FRSS). Making apparent to our readers these boundaries and the exclusions on which they depend requires that we carefully situate our version of FRSS. Toward this end, we first provide a brief overview of FSS. We then situate FSS and FRSS in relation to established scholarship in rhetorics of science as well as explicate key theories and concepts—posthumanism, feminist new materialism, posthumanist rhetorics, and feminist posthumanist rhetorics—on which these relations hinge. Doing so enables us to accomplish three interrelated goals: (*1*) to develop an origin story for FRSS that acknowledges its indebtedness to the rich body of scholarship that already exists within the field of rhetoric as well as extend that work; (*2*) to connect scholarship in the rhetoric of science to that in FSS; and (*3*) to identify undervalued feminist scholarship (both within and beyond rhetoric studies) that has much to contribute to posthumanist rhetorical theory.

Feminist Science Studies

> Social inequality shapes not only what science is done and how it is done . . . , but what science remains undone.
> —Nancy D. Campbell, "Reconstructing Science and Technology Studies"

Broadly defined, FSS examines the ways in which scientific theories emerge within specific historical, social, and economic contexts; challenges scientific research used to legitimate oppression and violence; and considers how categories of difference affect the making and meaning of scientific knowledge, and how these have led to exclusions, both in terms of who does science and what science does.[1] Histories that emphasize FSS's activist origins highlight the influence of the 1970s women's movement in the United States and note the particular importance of the book *Our Bodies, Ourselves*, which "gave rise to feminist self-help groups in many countries" (Lykke, 2008, p. 10). Another kind of activism in the 1970s was launched by feminists working in the sciences, such as Evelyn Fox Keller, who both "[critiqued] the ingrained phallogocentrism of the natural, biomedical, and technical sciences" (p. 10) and sought to improve the status of women scientists (Richardson, 2010; Subramaniam, 2009). Historicizing FSS via publications about it, Sarah S. Richardson (2010) identifies a 1978 special issue of *Signs* titled "Women, Science, and Society" as a groundbreaking moment (p. 339). This issue published substantive feminist critiques of science that became and remain major research threads in FSS.

In her history of FSS, Banu Subramaniam (2009) categorizes feminist critiques of science using six key themes. The first of these focuses on critiques of biological research, especially research on sex differences that supports theories of biological determinism (see also Richardson, 2010, pp. 342–333). These critiques specifically examine "how dominant cultural ideologies and normative structures find their way into scientific theories and laws" (Subramaniam, 2009, p. 956). The other five themes include reproduction and the labor of women, especially the increased "medicalization of women's reproductive capacities" (p. 957); discursive and visual representations of gender in scientific texts, particularly metaphors used in science (p. 958); the relationship between capitalism, colonialism, the appropriation of Indigenous knowledges, and scientific development (pp. 959–960); the challenging of boundaries between nature and culture (p. 958); and the production of scientific knowledge (p. 959).[2] This last theme is one with which feminist rhetoricians are familiar, for it includes contributions by feminist philosophers of science whose theories contest scientific claims to objectivity and value-free knowledge (p. 959; Richardson, 2010, p. 339). Included among these theories are Sandra Harding's (1992) "strong objectivity" and Donna Haraway's (1988) "situated knowledges." Indeed, contributions by these and other feminist philosophers of science lead Subramaniam (2009) to conclude that "if there were a discipline in which work in feminist science studies should be singled out for particular attention, it would be philosophy" (p. 960).

While FSS's investment in critiques of scientific representations and positivistic claims to knowledge resonates with rhetoric's commitment to analyses of language as well as its epistemic function, FSS's attention to material concerns, especially in relation to human embodiment, remains a firm reminder that rhetorical concerns exceed the symbolic. That discourse and matter form complex entanglements lacking distinct boundaries is captured most prominently in FSS by Haraway's figure of the cyborg: "partly human/animal and partly machine, partly organic body and partly technological artifact," the cyborg "is a figuration that signals a collapse of the central dichotomies on which positivist science is based" (Lykke, 2008, p. 12). Haraway's cyborg also marks the emergence in FSS of feminist technoscience, a field of inquiry that fuses biology and technology as it investigates scientific practices.

Following Bruno Latour, J. Blake Scott (2003) describes technoscience as the study of "a shifting ensemble of heterogeneous cultural practices rather than a discrete, coherent enterprise" (p. 229). *Feminist* technoscience continues this work but also considers the co-constitution of gender and technology and the complexities of "embodiment and the social and cultural situatedness of the scientist" (Lykke, 2008, p. 10; see also Wajcman, 2010). In so doing, it not only considers "the list of actors marked by gender, race, ethnicity, class, and sexual identity—i.e., human actors whose positions have been analyzed in different ways by feminist studies and cultural studies" but "may also include the perspectives of nonhuman actors" (Lykke, 2008, p. 12). This shift in focus—away from critiques of scientific knowledge via analyses of discourses produced by human beings to the tracking of scientific practices, which include but necessarily exceed *only* human actors—can be understood as the emergence of posthumanism in FSS.

Conceptual Hinges: Posthumanism, Feminist New Materialism

Far from being a monolithic and stable area of inquiry, posthumanism—or, more accurately, the various iterations of the most recent posthumanist turn—is as complex and enmeshed as the phenomena that give rise to it.[3] Most notably, these phenomena include new forms of biomedical and digital technologies that increasingly blur the assumed stable boundaries constitutive of Western Enlightenment's autonomous human subject. As N. Katherine Hayles (1999) explains,

> the posthuman implies not only a coupling with intelligent machines but a coupling so intense and multifaceted that it is no longer possible to

distinguish meaningfully between the biological organism and the informational circuits in which the organism is embedded. (p. 35)

This inescapable coupling gestures toward what Rosi Braidotti (2013) describes as a "relational ontology" (p. 190), which refuses the Cartesian hierarchy of mind over matter and thus challenges assumptions undergirding Western claims to human exceptionalism. Such assumptions include human animals' inherent superiority over—and thus ability and right to control—nonhuman animals, objects, and lands, and the "inferior" human animals associated with each. For this reason, posthumanism shares affinities with fields beyond science and technology studies, and posthumanist theories and methodologies emerge from and contribute to scholarship in areas such as critical animal studies, critical race studies, ecocritical/environmental studies, and postcolonial studies.[4]

Ultimately, contemporary posthumanist approaches recognize "that we are tied into complex systems or ecologies that decentralize the primacy of the human subject" (Trader, 2013, p. 204). Given this, posthumanist inquiries focus on networks of practices and webs of relations composed of elements in excess of the human. Two conclusions key to the development of posthumanist frameworks follow from this focus: (*1*) that we pay attention to how diverse elements interact—what they do and how they move, flow, circulate, and assemble; and (*2*) that we attend to the complex materiality of these elements in their dynamic interaction.

This second conclusion is one that has, according to Alaimo and Hekman (2008), "long been an extraordinarily volatile site for feminist theory" (p. 1)—and for good reason. Explaining how a focus on materiality risks reproducing the very assumptions feminist constructivist critiques of science sought to contest, Samantha Frost (2011) observes:

> From studies of the economic, imperial, and political forces that historically have shaped biological classifications of sex and race, to analyses of the ways in which political and cultural imperatives shape the movements of identification and desire, to explorations of the extent to which social and cultural practices transform bone and flesh, the insights and methods of constructivism have been crucial to feminist challenges to claims that import, encode, and at the same time deny power relations by presenting propositions as true or certain knowledge or as objective or natural fact. (p. 71)

Indeed, in our overview of FSS, we note the significant role critiques of biological determinism have played in the formation of the field. Recognizing the importance of these constructivist critiques, Elizabeth Grosz (2008) nevertheless argues that

> there is a certain absurdity in objecting to the notion of nature or biology itself if this is (even in part) what we are and will always be. If we *are* our biologies, then we need a complex and subtle account of that biology if it is to be able to more adequately explain the rich variability of social, cultural, and political life. (p. 24)[5]

Work by feminist materialists, including the scholars whose contributions appear in *Material Feminisms*, as well as others—Anne Fausto-Sterling (2000) for example—offers such complex accounts of human bodies, thereby "push[ing] our understanding of embodiment beyond the parameters of cultural and discursive construction. Indeed, each of them asks us to reckon with the materiality of a body that resists as well as conforms to cultural scripts" (Breu, 2014, p. 8). In this way, feminist materialism "refuses the linguistic paradigm, stressing instead the concrete yet complex materiality of bodies immersed in social relations of power" (Dolphijn & van der Tuin, 2012, p. 21).

More recently, scholars in feminist materialist cultural theory have continued to contest social constructivism's status as the only viable feminist response to biological determinism by extending the study of materiality beyond human bodies to include all forms of matter.[6] Informed by postclassical theories of matter, these scholars, often referred to as "new materialists," contest claims that material objects are stable, discrete entities that move in predictable ways in response to external forces. Instead, they regard the material world as active and "indeterminate, constantly forming and reforming in unexpected ways. . . . '[M]atter becomes' rather than that 'matter is'" (Coole & Frost, 2010, p. 10). Physicist Karen Barad (2007) similarly argues that "'matter' does not refer to an inherent, fixed property of abstract, independently existing objects; rather, *'matter' refers to phenomena in their ongoing materialization*" (p. 151). Yet in contending with matter as "vibrant, vital, energetic, [and] lively" (Bennett, 2010, p. 112), we need not render social, cultural, and discursive factors insignificant. Indeed, as political theorists Bonnie Washick and Elizabeth Wingrove (2015) explain, a key tenet of new materialism is that "the terms through which we know and name the world contribute directly to its materialization" (p. 65). Thus, "meaning and materiality are inseparable" (Simonsen, 2012, p. 15), and the challenge is to

> track the complex circuits at work whereby discursive and material forms are inextricable yet irreducible.

> ... For critical materialists, society is simultaneously materially real and socially constructed: our material lives are always culturally mediated, but they are not only cultural. (Coole & Frost, 2010, p. 27)

A key aim of new materialist inquiry, then, involves examining the agential interactions through which meanings, bodies, and objects are *co-constituted*.

New materialism as we have outlined it in this chapter—with its focus on tracking the dynamic interactions that co-constitute meaning and matter—thus offers one useful framework for scholars undertaking posthumanist inquiries. Indeed, this new materialist perspective is already informing innovative research in the field, most notably Laurie E. Gries's (2015) new materialist approach to visual rhetorics. Yet as contributors Kyle P. Vealey and Alex Layne discuss in the next chapter, not all theories gathered under the banner of "new materialism" are the same. Broad categorizations of new materialism can, for example, lump together the explicitly feminist projects of scholars such as Diana Coole and Samantha Frost, Elizabeth Grosz, and Karen Barad with that of Latour's Actor-Network Theory (ANT). As contributors Jen Talbot and Jennifer Bay explain in their respective chapters, ANT offers a useful methodology for tracking and describing scientific practices, but it can also pose a challenge to feminist theorists because its emphasis on "the symmetry between humans and nonhumans tends to obscure asymmetries between human actors" (Oudshoorn, Brouns, & van Oost, 2005, p. 103).[7] Yet these asymmetries remain crucially important to feminist scholars. Indeed, as Barad (2007) reminds us, "Haraway long ago emphasized that ... cyborg politics are not merely about the cross between human and machine but also about the technobiopolitics of the differentially human" (p. 59). Given its sustained commitment to theorizing human embodiment in its differentially valued and lived complexity, we argue that *feminist* new materialism offers an especially productive framework for scholars undertaking feminist posthumanist projects in the Rhetoric of Science, Technology, and Medicine (RSTM), a claim we believe is supported by the careful and thoughtful uptake of feminist new materialist theories that appears throughout this collection.[8]

Rhetorics of Science

> Materiality, as contrasted with symbolicity, occupies an increasingly prominent role in other humanities and social science disciplines, and rhetoricians of all stripes—especially rhetoricians

> of science, technology, and medicine—will benefit from making connections with these larger academic discourses.
>
> —Lisa Keränen, "Conspectus"

The posthumanist approach characteristic of contemporary feminist technoscience parallels an emerging shift in the rhetoric of science. In her introduction to a special issue of *Poroi* dedicated to charting trajectories for future research in RSTM, issue editor Lisa Keränen (2013) describes this shift in terms of a tension "between modernist, textualist approaches to RSTM and non- or even post-modern, post-human approaches that stress materiality and practice." Such a tension illuminates how RSTM, like FSS, is a diverse field of inquiry changing in response to emerging theories and technologies. Given the epistemic affordances of this diversity and the productive debates it can generate, we choose to introduce FRSS as one actor participating in these debates. That is, rather than pick a side and argue that it's the best approach for scholars doing RSTM work, we instead situate FRSS as offering *one* timely and rhetorically effective response to J. Blake Scott's (2003) call "for the development of new hybrid approaches that focus on different dimensions and functions of rhetoric" (p. 34). Scott makes this assessment as he outlines his own hybrid rhetorical-cultural approach, which follows his careful overview of three dominant approaches to research in the rhetoric of science. In order to situate both FSS and FRSS in relation to these various approaches, we summarize Scott's overview and then discuss how FRSS builds on and extends his hybrid approach.

The first group that Scott describes is composed of close rhetorical readings of individual texts written by "such scientific 'giants' as Darwin, Newton, Watson and Crick, and Robert Gallo" (p. 17). Citing work by Jeanne Fahnestock (1996), Alan Gross (1996), and S. Michael Halloran (1984) as representative examples, Scott explains that this approach is typically text- and agent-centered and "interprets the text in light of its immediate context, accounting for such factors as the writer's motives, the audience's values, and any rhetorical constraints imposed by the situation" (p. 17). Some work, such as Leah Ceccarelli's (2001), also "situates the text in a larger intertextual conversation," reading the text by examining its reception (via book reviews, editorials, and so forth) (Scott, 2003, p. 17). Because of its emphasis on textual analysis, this group of approaches aligns methodologically with the focus in FSS on text-based critiques of established biological research.

It is precisely this group's strength in methods of formal rhetorical analysis that can contribute to FSS's critiques of science. Thus, for example, while Banu

Subramaniam's overview of FSS summarized here acknowledges research that problematizes how metaphors work to create problematic gender representations in biological research, it does not recognize rhetorical analysis as a method of critique. Beyond analyzing metaphors as literary tropes, rhetorical critiques also consider their use and effects in relation to an author's motives, the values of the targeted audience, and the specific situational exigencies and constraints. In short, rhetorical critiques of science can help scholars in FSS better understand "the choices that rhetors make against a background of possibilities" as well as "the 'why behind' and 'what follows from' these choices" (Fahnestock, 2013).

The second group of approaches Scott identifies focuses on the production of scientific knowledge. In contrast to analyses of individual texts' rhetorical features, studies in this second group focus "on dynamic, social, and intertextual knowledge"; examples include "genre-discourse analyses" by scholars such as Greg Myers (1990) and Carol Berkenkotter and Thomas Huckin (1994) that "explain how networks of scientists socially (or sociocognitively) produce and negotiate scientific knowledge" (Scott, 2003, p. 18). Here, then, the method of genre-discourse analysis could extend FSS investigations of scientific knowledge beyond critiques of objectivity and value-neutral knowledge into a consideration of how and why *through language* community members are persuaded to accept some claims to truth and not others. Through such a framework, objectivity emerges as a rhetorical concept needing to be explained, as opposed to a static concept needing only to be critiqued.

Describing the third group as being concerned with the "public rhetoric of science," Scott (2003) explains that this kind of research, such as Celeste Condit's (1999) *Meanings of the Gene*, focuses on "rhetorical controversies *about* science and technology that are played out in different public forums" (p. 18). None of Subramaniam's categories explicitly recognize such controversies as a thematic in FSS research, yet the public understanding of science is a thriving multidisciplinary area of inquiry that has much to contribute to it, particularly with regard to understanding how "pseudo-controversies" are produced and sustained (Ceccarelli, qtd. in Condit, Lynch, & Winderman, 2012, p. 394). Additionally, FSS scholars investigating the relationship between capitalism and scientific development would benefit from rhetoric scholarship that examines how, for example, scientific knowledge circulating in public discourses reproduces neoliberal rhetorics of choice and self-improvement (see Thornton, 2011).

Clearly, the three groups of approaches Scott identifies have made and can continue to make important contributions to ongoing research in FSS. Yet despite the insights afforded by them, they are not without their limitations. The first group, for example, which focuses on close readings of individual texts, is limited by the tendency to reduce the complexities of context to "'influences' that help explain the success or failure of the text's rhetoric." Such explanations downplay the ways in which "extrarhetorical forces" (Scott, 2003, p. 18), such as corporate deployments of economic and political power, constitute the conditions of possibility for rhetorical action. The second group of approaches seems to correct for this limitation by enabling scholars to track how broader cultural movements and ideologies influence the making of scientific knowledge. Yet, according to Scott, they reinforce notions of knowledge production as a closed system by "limit[ing] their scope to the social practices of discrete scientific discourse communities" (p. 18). Furthermore, like the approaches described in the first group, these studies tend to offer explanations of how knowledge is negotiated rhetorically without also considering how rhetors might civically intervene in the production and application of unethical scientific knowledge (p. 18). And finally, while studies in the third group "seemingly expand the domain of science and the rhetorician's scope, they also often reinforce rather than problematize boundaries between the scientific and technical on the one hand and the social and cultural on the other" (p. 19).

In response to these limitations, Scott proposes what he calls a "hybrid rhetorical-cultural approach" that aligns with both Condit's (1999) theory of rhetorical formation and the posthumanist stance of feminist technoscience. Specifically, Scott (2003) outlines a methodology that enables rhetoricians to account for science's

> broader conditions of possibility . . . by treating [science] as a messy and dynamic enterprise and by tracking networks of influence wherever they extend. Often they extend across traditionally drawn boundaries of science, and often they involve nonhuman and extradiscursive "actors." (p. 22)

By situating individual scientific texts as only part of a larger network, Scott's methodology speaks back to the limitations of formal rhetorical analyses, which typically "follow the humanist paradigm in reading science through the strategic rhetorical moves of scientists" (p. 22). Furthermore, by enabling a more nuanced study of how human and nonhuman actors, relations of power, and broader cultural movements interact to produce scientific knowledge and generate different kinds of effects, Scott's methodology transgresses boundaries between scientific and social practices. It also moves beyond explanation as the primary goal of rhetorical inquiry. Instead, its goal is "to critique and work to intervene in problematic practices, oppressive power relations, and harmful effects. The main reason for troubling boundaries of science is to rebuild those boundaries in more ethical ways" (p. 21).

Scott's posthumanist methodology resonates with more recent scholarship in transnational technoscience, which examines the movement of scientific practices across multiple boundaries, including those of nation-state (see, for example, Prasad, 2014; Takeshita, 2011). It also contributes to research in RSTM that is less concerned with questions of epistemology (for instance, debates about bias and objectivity in the production of scientific knowledge) and more interested in mapping networks of scientific practices for purposes of intervening in them, a shift that Carl Herndl and Lauren Cutlip (2013) describe as moving "from a focus on saying and representing to a concern for doing and intervening, from 'how do we know?' to 'how can we act?'" In following the new materialist tenet that matter and meaning are inseparable, contributors to *Feminist Rhetorical Science Studies* attempt to do both: to couple ways of knowing with ways of being and acting in the world.

Conceptual Hinges: Posthumanist Rhetorics, Feminist Posthumanist Rhetorics

Since the publication of Scott's essential work, scholarship in rhetoric studies that is labeled explicitly "posthumanist" has gained significant traction. Reviewing its history of emergence in the field, Casey Boyle (2016) observes:

> Rhetoric has had its fair share of posthumanist moments. Writing over a decade ago, John Muckelbauer and Debra Hawhee argue that posthumanism compels us to engage "humans as distributed processes rather than as discrete entities" and that, in an age of posthumanism, "rhetoric becomes an art of connectivity and thereby asks for new considerations from multiple angles." (p. 539)

More recently, posthumanist approaches that understand rhetoric as "an art of connectivity" can be found in scholarship that theorizes rhetoric via concepts such as ecologies, networks, systems, and/or assemblages.[9] Common among these theories is an emphasis on movement. Synthesizing much of this scholarship, Chris Mays (2015) explains that

> thinking in terms of the movement—in the sense of the circulation, flow, interaction, and evolution—of elements in a rhetorical ecology can lead to a more fully developed theorization of the diachronic, dynamic, and—significantly—the *effect-laden* nature of our rhetorical (and non-rhetorical) ecosystems.

Key to posthumanist theories of rhetoric, then, is a focus on rhetoric as a dynamic phenomenon, one in which rhetorical elements (texts, purposes, audiences) move, interact with, change, and are affected by other agential elements within some larger system.[10] Accordingly, and in keeping with its combined commitments to decentering the autonomous human rhetor and attending to complex movement, a posthumanist rhetorical inquiry does not assume that a unidirectional causal relationship exists between a human rhetor's conscious purpose and the effects her intentional actions generate. Instead, posthumanist rhetoricians attempt to track interactions among elements, which means they focus on what elements, human and nonhuman, are *doing* and how repeated doings, or practices, interact such that some things happen and others don't.[11]

Implications for rhetoric that arise as a consequence of such assumptions include a rethinking of rhetorical agency as rhetorical action, with action itself understood as interdependent *inter*action. Furthermore, in acknowledging that human beings alone cannot and do not make things happen, a posthumanist perspective conceptualizes rhetorical effects as both contingent and emergent: although what happens can neither be predicted nor controlled, it *can* happen otherwise. As such, a posthumanist rhetorical approach focuses less on how humans can assert our individual wills to solve problems and more on how we can contribute to "the articulation or assemblage of the conditions for the *emergence* of solutions" (Hawk, 2011, p. 90). Importantly, then, the fact that rhetorical effects emerge via interactions that can include but necessarily exceed human intentions does not prevent humans from working *with* intention to bring about future change: we can participate in inventing practices through which elements might come to be assembled *otherwise*.

Because they attend carefully to how meanings, bodies, and objects interact, posthumanist rhetorical theories, like posthumanist theories more broadly, are materialist in orientation. This materialist bent, however, differs from more traditional theories of material rhetoric in significant ways. In his review of material rhetorics, J. David Cisneros (2008) identifies two major strands: rhetoric as material and matter as rhetorical. In the first category he situates influential early work by scholars in communication studies, such as Michael Calvin McGee (1982), Celeste Condit (1994), Lisa Flores (1996), and Ronald Walter Greene (1998). Drawing specifically on Greene, Cisneros (2008) explains that this strand examines "the ways in which rhetoric contributes to the material conditions of existence, particularly the relationship between individuals and institutions"; in particular, it considers how "discourse constructs material relations of power, legitimizes ways of being, and is productive of effects." Accordingly, this strand conceptualizes rhetorical inquiry as being primarily concerned with understanding how humans' uses of symbols (alphabetic texts and images for instance) generate material effects—how, for example, language is used to control bodies, privileging some while doing violence to others, thereby sustaining asymmetrical relations of power.[12] But humans can also use symbols to challenge these asymmetries. As Jamie White-Farnham (2013) argues in a review of more recent scholarship in material rhetorics, understanding rhetoric as material involves not only analyzing the ways in which discourses produce material injustices but also *producing* discourses that try to rectify them. On this view, then, material rhetorics refer to "symbolic action[s that are] deployable in contexts of struggle, whether loosely defined social milieu or formalized institutions," with the goal being "to effect social and material change" (p. 475).

Drawing on Carole Blair's important contribution to Jack Selzer and Sharon Crowley's *Rhetorical Bodies* (1999), Cisneros (2008) explains that the second strand—matter as rhetorical—does not begin with a focus on how humans' uses of symbols affect their material lives. Instead, it considers how texts, broadly defined to include "all kinds of things, from bodies to monuments, participate in discursively shaping our reality." Specifically, this approach examines "what the text does in its materiality," how a thing *in its very materiality* persuades (Cisneros, 2008). In Cisneros's own analysis of La Gran Marcha, a 2006 social protest for immigrants' rights, he demonstrates the affordances of this second approach. Rather than rhetorically analyze protestors' signs or the lyrics of music streaming through the crowds, he instead theorizes the ways in which protestors' material bodies collectively signified a revised immigrant body, one that was both publically visible and politically aware.

While we acknowledge the enormous contributions these two dominant approaches to material rhetorics have made and continue to make to the field, we also want to value what posthumanist rhetorics bring to the conversation: frameworks for theorizing rhetoric in terms of interactions between meaning and matter wherein *nonhuman* elements actively participate alongside symbol-using animals to effect change. On this view, then, nonhuman objects are neither "extratextual" (White-Farnham, 2013, p. 479) nor "extrarhetorical" (Scott, 2003, p. 18); rather, they function as rhetorical actants in their own right, exerting an agential force that exceeds their signification to and for humans.[13] Posthumanist rhetorics thus continue the conversation in material rhetorics begun a decade ago by scholars who theorize the rhetoricity of nonhuman objects in their dynamic interactions (see, for example, Helmers, 2006; Marback, 2008). They also extend earlier scholarship that contends with both "the discursivity of material action and the materiality of discourse" (Hesford, 1999, p. 209), thereby blurring the boundary between language and matter. Particularly noteworthy in this regard is Kristie Fleckenstein's (1999) theory of somatic mind, which conceptualizes identity formation as "a material-discursive fusion" (p. 298) that involves nonhuman elements in the surround. Finally, with respect to more recent theories of material rhetoric and this collection specifically, *feminist* posthumanist theories of rhetoric complement contributions to Barbara Biesecker and John Lucaites's *Rhetoric, Materiality, and Politics* (2009), which, while expanding our understanding of material rhetorics through the use of Foucauldian, Derridean, Marxist, and psychoanalytic frameworks, do not engage scholarship in material feminisms.

Given that posthumanist rhetorical theories emphasize dynamic interactions between meaning and matter, it's perhaps no surprise that work by Barad is emerging as being especially important, particularly for scholars undertaking feminist posthumanist projects in RSTM, such as recent work by Christa Teston (2016) and Molly Margaret Kessler (2016) as well as many of the chapters gathered in this collection.[14] Barad defines posthumanism as an investigation of "the material-discursive boundary-making practices that produce 'objects' and 'subjects' and other differences out of, and in terms of, a changing [dynamic] relationality" (2007, p. 93). For feminist rhetoricians, Barad's version of posthumanism proves useful because it enables us not only to examine how differences within the category of human get made and to what effect but also to recognize the production of such differences as involving more than only the use of language. Thus, while a humanist rhetorical approach might posit social inequality as an effect of human agents using symbols to legitimate and sustain oppressive ideologies, a feminist posthumanist approach would conceptualize oppressive practices as material-discursive relations that produce distinctions undergirding asymmetrical relations of power. Furthermore, unlike posthumanist theories that in privileging the agency of nonhuman matter ignore the force of language or the momentum of dominant ideologies, Barad's posthumanism enables feminist rhetoricians to tarry with the material without dismissing or diminishing the significance of the social. Another way to think of Barad's emerging importance to feminist posthumanist theories of rhetoric is by way of this analogy: Barad is to feminist posthumanism as Latour is to posthumanism.

Particularly essential to projects in feminist new materialist rhetorics are key Baradian concepts that appear throughout this collection, such as *material-discursive* and *intra-action*. The first combines and then revises Donna

Haraway's concept of *material-semiotic* with Judith Butler's theory of performativity to develop a theory of posthuman performativity that recognizes the agential force of nonhuman matter.[15] For Barad, discourse does not refer to words or symbols; rather, discursive practices are specific arrangements of the material world that create distinctions and thus "[constrain] and [enable] what can be said. Discursive practices define what counts as meaningful statements" (2003, p. 819). Enactments of border fences between nations, for example, are discursive practices that enable and constrain the intelligibility of "citizen." Barad also challenges dominant assumptions about matter when she explains that "*matter is . . . not a thing, but a doing,*" an interdependent process that manifests a "*congealing of agency*"—human and nonhuman (2003, p. 822). Returning to the example of the border fence: matter refers not to the material fence itself as a static thing; rather, it's about how the fence's chemical composition, the geology of the land on which it is situated, the bodies of those who build it as well as those whose movements are constrained by it, and the processes involved in its being imagined, debated, built, and maintained *interact* such that a particular border fence comes into being. Simply put, Baradian matter is about matter's *materialization*. Taken together, *material-discursive* refers to "*what it means to matter*" (2003, p. 824). That a specific border fence exists manifests a particular kind of material-discursive entanglement, one that helps us understand, with complexity, how and why some bodies have become intelligible as citizens while others have not. For Barad, our ethical responsibility resides in the realization that we participate in the making of this difference: we must recognize how "the world is materialized differently through different practices," which requires that we take "responsibility for the fact that our practices matter" (2007, p. 89).

The second key Baradian concept that several of our contributors engage is *intra-action*. Although posthumanist rhetorics examine carefully interactions between meaning and matter, new materialists seek to understand their co-constitution. As such, from a new materialist perspective, *interaction* is an insufficient concept for theorizing rhetorical movement and change, since it "presumes the prior existence of independently determinate entities" (Barad, 2008, p. 170). In contrast, Barad's intra-action "*signifies the mutual constitution of entangled agencies*" and thereby "recognizes that distinct agencies do not precede, but rather emerge through, their intra-action" (2007, p. 33). On this view, the power to delimit citizen from noncitizen does not belong to the border fence or to the officers who patrol it; rather, this capacity emerges from and is enacted through a complex process of materialization that could have happened *and can still happen* otherwise. Intra-actions (what we refer to as "material-discursive entanglements" earlier) thus produce distinctions that are contingent and ongoing, not essential or determined. Importantly, this contingency manifests moments of becoming that are also spaces of intervention. As Laurie Gries (2015) explains,

> becoming is an opening up of events into an unknown future. Reality is change, an open process of mattering and assemblage. From such perspectives, a new materialist rhetorical approach recognizes that things constantly exist in a dynamic state of flux and are productive of change, time, and space. (p. 86)

Understanding how some changes happen to the exclusion of others is rhetorical work: by engaging as rhetoricians in the world's continual becoming, we can participate in remaking boundaries and meanings of difference by helping to enact alternative material-discursive entanglements.

While we acknowledge that this discussion of posthumanist rhetorical theories has taken us far afield from understanding rhetoric as the human art of persuasion via the strategic use of symbols, we also believe there is much to be gained—for both the field and the world beyond—if we imagine rhetoric as participating in the much larger project of trying to understand how change happens. The risk, of course, is that in joining such a project we become unable to define rhetoric as a distinct field of inquiry. From a new materialist perspective, however, this inability might be more productively thought of as an inevitability: no field of inquiry is distinct. *Feminist Rhetorical Science Studies* acknowledges this inseparability. Indeed, it emerges from it. Yet as we believe the chapters that follow demonstrate, feminist posthumanist approaches to science that expand our understanding of what rhetoric is and can become have much to contribute, and they can do so without sacrificing the knowledge-making practices that render them distinctively rhetorical.

The Chapters

Each of the chapters that follow responds to a key question that informed our editorial selections: how can feminist rhetoricians of science engage productively and responsibly with emerging theories of the posthuman? Specifically, each chapter offers a feminist rhetorical approach to science studies capable of forging innovative, interdisciplinary connections between feminist rhetorics, posthumanist rhetorics, rhetorics of science, and FSS. Significantly, these approaches are diverse and at times conflicting; as such, the collection as a whole invites productive debate that we hope informs future research within and beyond rhetoric studies.

One key debate that came to the fore as we worked with the chapters concerns whether and how existing posthumanist frameworks such as ANT and object-oriented ontology (OOO) can contribute to *feminist* rhetorical science studies. As our own earlier discussion suggests, we believe FRSS scholars need to engage with these frameworks cautiously, if at all. In order to more fully develop these concerns, as well as to help our readers understand why and how the theoretical and methodological choices we make as scholars quite literally matter, we begin with Kyle P. Vealey and Alex Layne's "Of Complexity and Caution: Feminism, Object-Oriented Ontology, and the Practices of Scholarly Work." In their chapter, Vealey and Layne identify important differences in epistemological assumptions between two categories of posthumanist scholarship—OOO and feminist new materialism. As we note in our prologue, these two categories of inquiry are frequently lumped together under various headings (for example, posthumanist studies, speculative realism, new materialisms, object-oriented philosophy, and the material turn), which elides significant differences between them. Vealey and Layne argue that as feminist researchers we must contend with these differences, since an uninformed uptake of OOO can work against a key goal of feminist rhetorics: namely, to attend to issues of difference in the pursuit of what might be termed material justice. That our scholarly practices "constitute, rather than reflect, ontologies" is one of the authors' key points, the stakes of which are clearly demonstrated in their examination of how OOO emerges as its own kind of object via its exclusion of feminist thought.

Through their examination, the contributors highlight how influential scholars of OOO, in theorizing the autonomy of objects, turn attention away from examining how intra-actions constitute different kinds of matter that matter differently to differentially embodied human beings. Vealey and Layne thus remind us that methodologies are material knowledge-making practices: they make selections that acknowledge some histories to the exclusion of others and, in so doing, help bring into being some worlds but not others. As such, it is incumbent of us to exercise caution and care as we adopt and develop frameworks for doing research in FRSS.

The remainder of the chapters, we believe, model this kind of methodological care, demonstrating how feminist scholars can engage posthumanist frameworks rhetorically while also—or rather, precisely by—foregrounding the political and ethical stakes involved in doing so.

In chapter 2, "Flat Ontologies and Everyday Feminisms: Revisiting Personhood and Fetal Ultrasound Imaging," Jen Talbot uses legislation that posits the existence of fetal personhood as a scene for examining how Latour's

ANT and Barad's agential realism constitute motherhood and maternal accountability differently. Talbot begins by explaining how the posthumanist concept of nonhuman agency poses a threat to women's reproductive rights. She specifically examines the role of fetal ultrasound technology in constituting networks of agents that participate in constructing fetal personhood. An ANT approach, she explains, ignores "the fetus's biochemical influence on the body," thereby constructing the pregnant woman and fetus as distinct and symmetrical agents in the same network. Such a construction, she continues, problematically supports antiabortion legislation that claims to speak for the "autonomous but silenced subject."

Because of ANT's capacity to undermine women's reproductive freedom, Talbot advocates instead for Barad's agential realism, which theorizes nonhuman agency in a way that does not construct the pregnant woman and fetus as opposing and distinct agents. For Barad, phenomena (entanglements of discursive-material intra-actions), rather than individual bodies, are the primary ontological units through which subjectivity *emerges*; as such, the fetus as subject can be understood as an effect of intra-actions to which those who participated in its emergence are accountable. As Talbot explains, within this framework

> women are accountable not to the fetus as person nor to the state or the public that would police her adopting the role of mother. Rather, those who would construct fetal personhood are accountable for the consequences of that articulation.

Among other things, this means that "a state that ascribes personhood to a fetus is accountable for providing the infrastructure of care for that fetus." In this way, Talbot demonstrates how agential realism's concept of nonhuman agency is capable of sponsoring interventions that are more in line with feminist political commitments. Along the way, Talbot also exposes what ANT ignores: the ways in which "agential force comes from the aggregation of chemical, neural, and electrical responses that make up human connection and manifest as social behaviors." In so doing, she evidences once again the value of a key feminist rhetorical tactic: identifying exclusions for purposes of contesting and revising dominant theories.

Further demonstrating the interdisciplinary reach of Barad's theory of agential realism is Catherine Gouge's "'The Inconvenience of Meeting You': Rereading Non/Compliance, Enabling Care," which extends the practice of ontologically integrating materiality and discursivity to the issue of patient care and the label of "noncompliance." Gouge argues that the construction of patients as noncompliant emerges from a biomedical perspective that separates the discursive from the material and erroneously demands patients follow a narrow path to "health"; any deviation from that path results in the label of "noncompliance." Challenging the ableist structures (informed by disability studies) and the "culture of compliance in biomedicine," Gouge argues for "more ethical, effective, and efficient systems of care." She reembodies so-called noncompliant patients as complex agents who make difficult choices about their health care based on their experiences not only of their diseases but also of their entire lives. Gouge notes some of the implications this piece has for FRSS, first by her response to calls for decentering the discursive and bringing feminist analysis to biology, and second by extending work in embodied rhetorics "beyond conventional humanist accounts of persuasiveness and influence." She posits her most significant implication as a "revaluing of noncompliance," which does not "assum[e] noncompliant acts are self-destructive signs of *failures* to cope." Instead, Gouge argues that we ought to regard noncompliant acts as

> emergent responses to complex, rhetorical situations in an individual's contexts for care. And, as such, they are actually *evidence of* coping. As health-care participants diverge and converge with prescribed or agreed-on therapies, the variables associated with our specific situations and our experiences of them change. Recognizing this prior to judgment and assigning responsibility is central to improving therapeutic alliances and systems of care.

This resonates with several chapters in this collection, and particularly with Talbot's and Liz Barr's attention to ethical care toward patients' bodies and lives. We see Gouge's work making important contributions both within and beyond our field, extending to medical practice as well.

Taken together, Talbot's and Gouge's chapters persuasively demonstrate how Barad's approach to feminist new materialisms can complicate and extend theories of rhetorical agency circulating in legal and medical rhetorics. They also model how posthumanist notions of rhetorical agency need not depoliticize feminist rhetorics. Indeed, their respective chapters make clear that a key contribution of feminist posthumanist rhetorical approaches is their ability to rethink agency in ways that lead to more ethical human relations.

In "Mattering Gender: Technical Communication and Human Materiality," Jennifer Bay extends this feminist posthumanist rhetorical project in her efforts to develop better methodologies for researching the embodied experiences of female technical communicators. Toward this end, Bay traces how connections across three levels of scale—the classroom, the workplace, and the field of Technical Communication—emerge differently depending on the researcher's selected methodological framework. She demonstrates this process by describing specific research projects. The first investigated whether a relationship existed between female students' experiences in professional writing majors and female practitioners' experiences in the workplace. Using a conventional methodology, Bay and her coresearcher were unable to find much data on the gendered experiences of women in the tech workforce. Bay explains this absence of data as an effect of her selected methodological framework's presumed objectivity, which could not perceive the material experiences the researchers sought to explain. Bay then describes and contrasts two alternative frameworks, each of which recognizes materiality as relations between bodies and technological objects. Her descriptions clearly demonstrate the new materialist tenet that *how* researchers study a phenomenon helps constitute the phenomenon they seek to understand. For Bay, feminist new materialisms offer more ethical methodologies for investigating the lived experiences of students and practitioners. Furthermore, and returning to her motivating pedagogical concern, Bay posits that "a feminist new materialist approach in which we teach students to see... technologies as cocreators of the world and as agential elements in any creative situation could provide reserved students with more confidence." In this way, Bay invites her readers to understand feminist new materialisms as research methodologies that are also pedagogies, ways of building new worlds that "make gender matter and make the mattering of gender show up for us differently."

The aim of making "gender matter and the mattering of gender show up differently" in the space of the laboratory is pursued in Jordynn Jack's "How Good Brain Science Gets That Way: Reclaiming the Scientific Study of Sexed and Gendered Brains." Jack argues for an engagement with neuroscience that moves us beyond critique or simple acceptance or rejection of neuroscientific findings—positions that are problematic as they can occur only *after* the production of scientific knowledge. Jack builds on works by Barad, Sandra Harding, and Haraway that explore the rhetorical nature of science, and she demonstrates how rhetoricians and neuroscientists might productively cross disciplinary boundaries. As Jack explains,

> only by understanding how scientific knowledge is produced through specific experimental entanglements can we begin to identify new possibilities that better capture the interaction of the sociocultural and neurobiological networks through which brain sex/gender differences may (or may not) emerge.

These entanglements invite rhetoricians to engage *with* neuroscience, a field that is itself interdisciplinary (including fields like psychology, sociology, physics, and biology), particularly in light of studies on gender and gender stereotypes. In order to explore such (potential) entanglements, Jack considers several studies in neuroscience and gender. Her analysis reveals that such studies often begin with "the assumption that sex/gender differences exist." These kinds of beginnings generate hypotheses, methods, and materials that then also conform to gender stereotypes, thus creating conformational results (for instance, that women are more empathic than men). Studies that emerge from a different ground—that question whether, why, and in what conditions sex/gender difference exist—often produce different results that do *not* conform to gender stereotypes. Jack's work contributes much to this volume, perhaps most specifically by demonstrating ways that rhetoricians can "analyze and disrupt" stereotypical gender representations, as well as by creating "a renewed impetus . . . to engage with brain science at the level of the experiment and research article." Her approach can extend this "renewed impetus to engage" to other fields, particularly scientific and medical, but beyond those boundaries as well.

The next two chapters in the collection—Daniel J. Card, Molly M. Kessler, and S. Scott Graham's "Representing without Representation: A Feminist New Materialist Exploration of Federal Pharmaceutical Policy" and Liz Barr's "Embodied Vernacularity at the FDA: Feminism, Epistemic Authority, and Biomedical Activism"—both examine representation and testimony through U.S. Food and Drug Administration (FDA) hearings, although the authors take quite different approaches and thus provide different ways of thinking through "representation" itself. Card et al. consider representation in FDA trials from an ontological perspective that differs from Barr's epistemological emphasis, as well as from several others in this collection. Specifically, they ground their study in Annemarie Mol's work on the body and disease (2002), highlighting the "argument that NM [new materialism] requires a transition from a (post-)modern politics of *who* to a politics of *what*." Here, the "who" represents a "'perspectival' approach to the body," wherein the body itself is left silent and "people with perspectives" representatively speak for it. The "what," then, is an ontological position that "foregrounds the *practices* through which disease comes into being—enactments of disease(s)," which results in multiple ontologies and diseases. Card et al. employ this who/what distinction in an evaluation of the FDA's patient representative program with the goal of establishing a more ethical approach to how patients are represented in Drug Advisory Committee meetings. After arguing for the value of a "what" approach, the authors then describe how they coded actual discourse from Drug Advisory Committee transcripts based on four ontological categories—lab, home, clinic, and market—that emerged from various practices described in the transcripts. This particularly rich section of the chapter offers extensive examples and also models an analytical approach that could be productively applied to future studies in FRSS. Although intersectional aspects of material bodies (such as sex, race, disability) are not addressed in the analysis, we see the authors laying the groundwork here for important future inquiry.

Card et al.'s chapter is also significant because it returns us to a key debate we identified at the start of this section: namely, how can, or should, existing posthumanist frameworks such as ANT and OOO contribute to *feminist* rhetorical science studies? Several of our earlier contributors tackle this debate by delineating distinctions between ANT, Activity Theory, and feminist new materialisms. In Card et al.'s chapter, however, a different cut is made. Specifically, in order to foreground the who/what distinction that is crucial to their argument, they gather scholarship that supports the "what" approach—speculative realism, OOO, and alien phenomenology, as well as work by Latour—under Coole and Frost's articulation of "new materialisms," which functions as "a convenient catchall term for these recent intellectual efforts devoted to making objects, things, matter, and the concrete central to inquiry." They then explain how this grouping of new materialism, which fundamentally rejects inquiry into epistemology and representation, opposes postmodernism and its preoccupation with the social/cultural/discursive constitution of reality, or the "who" approach.

Following Barad, we recognize these distinctions (who/what, postmodernism/new materialism) as emerging from "cuts," or boundary-making practices, that generate effects, one of which is the possibility that significant differences between theories gathered under the unifying banner of new materialism might be lost. As Vealey and Layne argue in their chapter, scholars in FRSS ignore these differences at our peril. Card et al. gesture toward this same concern when they state that "some variants [of new materialism] better lend themselves to feminist rhetorical scholarship than others," which they then persuasively demonstrate in their use of Mol's work. Another possible effect is that important continuities between postmodernism and new materialism might be ignored. Indeed, we disagree with the claim that postmodernism is concerned *only* with discourse, epistemology, and representation, and not concerned with the material.[16] Despite this disagreement—or, rather, because of it—we've included this chapter in the collection not only because it carefully explains and then models a complex analytical approach for doing qualitative research in FRSS but also because it can potentially spark dissention and productive debate in and beyond FRSS.

In "Embodied Vernacularity," Liz Barr takes an epistemological approach as she tackles the issue of community testimony in an FDA hearing for a new drug to be used as pre-exposure prophylaxis against HIV. After developing a theory of "embodied vernacularity," or "discourse that positions itself against an institution through a rhetoric of the body," Barr applies it in a close textual analysis of a transcript for the 2012 FDA advisory committee hearing on the drug Truvada. She contrasts the empirically based testimony of scientific and medical experts against the materially located testimonies offered by community members involved in the drug trials. She then considers how these different ways of knowing could optimally be used in a both/and approach, wherein the statistical and the embodied have (at least) equal persuasive powers. Although the FDA ultimately approved Truvada, and thus "embodied vernacularity may have proven less persuasive than institutionalized scientific discourse," Barr argues that community members' "embodied vernacularity . . . was an ethical act, one that sought to privilege the lived experiences of people with HIV." This reflects Gouge's call for a more ethical approach to patient compliance questions, again through embodying patients and empowering them to speak their own truths to power. And, as is the case with all of the previously introduced chapters, Barr's focus on the ethics of materiality reminds us of its vital importance in FRSS.

Like Card et al.'s chapter, we see Barr's work as capable of motivating productive dissention among scholars in posthumanist rhetorics, although for a decidedly different reason. Specifically, Barr's epistemological approach disrupts the posthumanist call to privilege ontology over epistemology as a strategy for debunking the centrality of the exceptional human knower. By examining the ways in which bodies, scientific data, and public testimonies interact and materialize in specific, power-laden ways, Barr shows how the exceptional human knower is a generic construct that occludes the processes through which some kinds of human knowing and being come to matter more than others. In so doing, she persuasively demonstrates that rhetoric scholarship can remain very much concerned with human symbolic action as it simultaneously contributes to the posthumanist project of collapsing the boundary between language and matter.

The final chapter, "Becoming-Thinking Otherwise, Rhetorically," considers the ways in which research in feminist rhetorics can contribute to feminist posthumanist studies across the disciplines. Specifically, the chapter demonstrates how distinctively rhetorical investigations of writing, research, and activist practices can motivate new ways of thinking about meaning-matter entanglements. By engaging in this kind of intellectual reciprocity, we aim to make rhetoric scholarship more accessible to scholars from other fields, striving to call into being cross-disciplinary feminist alliances that reverberate throughout and beyond the walls of academe.

Notes

1. Our introduction to FSS is designed to highlight dis/connections between it and scholarship in the rhetoric of science. It is not intended to offer a substantive overview of scholarship in or the history of FSS. Readers looking for such an overview might choose to begin by looking at texts frequently recognized in histories of FSS as being foundational to its formation. See, for example, Anne Fausto-Sterling's *Sexing the Body* (2000); Donna Haraway's *Simians, Cyborgs, and Women* (1991, which contains her essay "A Cyborg Manifesto"); Sandra Harding's *The Science Question in Feminism* (1986) and *Whose Science? Whose Knowledge?* (1991); Evelyn Fox Keller's *Reflections on Gender and Science* (1985); Helen Longino's *Science as Social Knowledge* (1990); Dorothy Roberts's *Killing the Black Body* (1997); and Nancy Tuana's *Feminism and Science* (1989). For more recently published readers in FSS, see Harding, 2011; Smelik and Lykke, 2008; Wyer, Babercheck, Cookmeyer, Öztürk, and Wayne, 2014.

2. For more on the relationship between colonialism, Western imperialism, and science, see Spurr, 1993, p. 168; Takeshita, 2011, pp. 35–36, 52.

3. For a review of different kinds of contemporary posthumanist theory, see Sharon, 2014.

4. Particularly noteworthy are projects that establish linkages across these fields, such as Mel Y. Chen's *Animacies: Biopolitics, Racial Mattering, and Queer Affect* (2012) and Alexander G. Weheliye's *Habeas Viscus: Racializing Assemblages, Biopolitics, and Black Feminist Theories of the Human* (2014).

5. For a similar argument within rhetoric studies, see Celeste Condit (2004), where she argues:

> The fact that humans must always interpret the world that is other-than-language through language does not mean that there is nothing other than language. Indeed, most of our experiences in life speak to the existence and material force of phenomena that are other-than-language (as in, for example, our experiences of pain, food, heat, or tornadoes). (p. 7)

6. See, for example, Ahmed, 2010; Alaimo, 2010; Barad, 2007, 2008; Chen, 2012; Coole and Frost, 2010; Grosz, 2008, 2010; Hekman, 2010.

7. Indeed, to deny the existence of such asymmetries prior to the description of a network, and further, to fail to include its various instantiations in that network, is antithetical to the very project of science and technology studies (STS): to "focus on science as an *achievement* that develops through local, contingent, negotiated, consensus-building, collective claims-making practices" (Whelan, 2001, p. 546). As sociologist Emma Whelan (2001) explains in her summary of and response to the methodology of apolitical description adopted by many STS scholars,

> their general point is that entities are what they are through heterogeneous networks of association; Pasteur-the-great-scientist or Andrew-the-businessman are the result not of the men-in-themselves but of the network of allies that support them—material objects, people, texts, organizations, and so on. Fair enough. But why can't racism be part of a heterogeneous network . . . ? Couldn't we imagine racism—or sexism or feminism—as elements of heterogeneous networks or networks in themselves, which support or undermine other networks? A staunch refusal to consider that oppressive ideologies may be part of a network amounts to assuming the oppressive ideologies are *not* part of the network . . . In effect, mainstream STSers blackbox questions of social inequality when they refuse to open them to scrutiny. (p. 559)

8. In advocating that posthumanist inquiries in RSTM engage with feminist new materialisms, we do not mean to suggest that other posthumanist theories and methodologies *can't* inform research in FRSS. For excellent examples that demonstrate how ANT can be responsibly deployed in the service of feminist rhetorical historiographies of science, see Hallenbeck, 2012, 2015.

9. For a discussion of rhetorical ecologies, see, for example, Brooke, 2009; Edbauer, 2005; Rivers and Weber, 2011; Shepley, 2013; for rhetoric as complex system, see Cooper, 2011; Jung, 2014; Mays, 2015; for rhetoric as network, see Blakesley and Rickert, 2004; Condit, 1996; Dingo, 2012; Gries, 2015; Lynch and Rivers, 2015; Rice, 2006; Ridolfo and Hart-Davidson, 2015; for rhetoric as assemblage, see Grabill, 2014; Hawk, 2011; Holmes, 2013; Nicotra, 2016. For a critique of the ways in which a posthumanist metaphorizing of "ecology" can dematerialize actual land ecologies, see Ortoleva, 2013; Ríos, 2015.

10. See, for example, Sidney Dobrin's collection *Writing Posthumanism, Posthuman Writing* (2015), which acts as "an attempt to incite and disrupt Writing Studies from the constraints of humanist thought" (p. 17). Although the collection is situated within the field of writing studies, its chapters range from emphases in animal studies to zombies to neurorhetorics and thus intersect with posthumanist theories of rhetoric more broadly.

11. For a retheorizing of "practice" in rhetoric and writing via a posthumanist orientation, see Boyle, 2016.

12. Other notable works that demonstrate the crucially important interventions this strand of rhetoric can make in the world include James Wilson and Cynthia Lewiecki-Wilson's *Embodied Rhetorics* (2001) and Davi Thornton's *Brain Culture* (2011).

13. One recent thread of posthuman rhetorics that explicitly considers the agential force of matter beyond its relation to humans is "object-oriented rhetoric," a term first used by Scot Barnett (2010) to describe Graham Harman's articulation of metaphor in his uptake of Martin Heidegger in object-oriented philosophy. A few examples of this include Thomas Rickert's *Ambient Rhetoric* (2013); James J. Brown Jr. and Nathaniel Rivers's "Composing the Carpenter's Workshop" (2013); Byron Hawk, Chris Lindgren, and Andrew Mara's "Utopian Laptop Initiatives" (2015); and Scot Barnett and Casey Boyle's *Rhetoric, through Everyday Things* (2016). A particularly rich subset of scholarship in object-oriented rhetoric is Latourian rhetoric. See, for instance, Paul Lynch and Nathaniel Rivers's *Thinking with Bruno Latour in Rhetoric and Composition* (2015).

14. Indeed, in her review essay "Material Rhetorics Meet Material Feminisms," Marita Gronnvoll (2013) forecasts this importance, observing that Barad "provides an intricate retheorizing of materialism that could be useful for extending material rhetoric. Seeing rhetoric through Barad's lens invites a paradigm shift that urges the critic to question the entire notion of rhetoric as object of study" (p. 106).

15. Barad explains the limits of Butler's theory of performativity for posthumanist projects as follows:

> Butler's theory ultimately reinscribes matter as a passive product of discursive practices rather than as an active agent participating in the very process of materialization. . . . Furthermore, Butler's theory of materiality is limited to an account of the materialization of human bodies or, more accurately, to the construction of the contours of the human body. Agential realism's relational ontology enables a further reworking of the notion of materialization that acknowledges the existence of important linkages between discursive practices and material phenomena without the anthropocentric limitations of Butler's theory. (2003, pp. 821–822, n26)

16. Michel Foucault's genealogical methodology, for instance, particularly as demonstrated in *Discipline and Punish* (1977) and *The History of Sexuality: Vol. 1* (1990), significantly engages with the construction and critique of discourse *and* practice, and his work on the body is dependent on the materiality of

training and discipline on the body itself. Butler (1993) also struggles (though some would argue unsuccessfully) with the materiality of the body in her book *Bodies That Matter*.

References

Ahmed, S. (2010). Orientations matter. In D. Coole & S. Frost (Eds.), *New materialisms: Ontology, agency, and politics* (pp. 234–257). Durham, NC: Duke University Press.

Alaimo, S. (2010). *Bodily natures: Science, environment, and the material self.* Bloomington: Indiana University Press.

Alaimo, S., & Hekman, S. (Eds.). (2008). *Material feminisms.* Bloomington: Indiana University Press.

Barad, K. (2003). Posthumanist performativity: Toward an understanding of how matter comes to matter. *Signs, 28*(3), 801–831.

Barad, K. (2007). *Meeting the universe halfway: Quantum physics and the entanglement of matter and meaning.* Durham, NC: Duke University Press.

Barad, K. (2008). Living in a posthumanist material world: Lessons from Schrödinger's cat. In A. Smelik & N. Lykke (Eds.), *Bits of life: Feminism at the intersections of media, bioscience, and technology* (pp. 165–176). Seattle: University of Washington Press.

Barnett, S. (2010). Toward an object-oriented rhetoric [Review of the books *Tool-being: Heidegger and the metaphysics of objects* and *Guerrilla metaphysics: Phenomenology and the carpentry of things*, by Graham Harman]. *Enculturation, 7.* Accessed 8 June 2016. Retrieved from http://enculturation.net/toward-an-object-oriented-rhetoric

Barnett, S., & Boyle, C. (Eds.). (2016). *Rhetoric, through everyday things.* Tuscaloosa: University of Alabama Press.

Bennett, J. (2010). *Vibrant matter: A political ecology of things.* Durham, NC: Duke University Press.

Berkenkotter, C., & Huckin, T. N. (1994). *Genre knowledge in disciplinary communication: Cognition/culture/power.* Mahwah, NJ: Lawrence Erlbaum Associates.

Biesecker, B. A., & Lucaites, J. L. (Eds.). (2009). *Rhetoric, materiality, and politics.* New York, NY: Peter Lang.

Blair, C. (1999). Contemporary U.S. memorial sites as exemplars of rhetoric's materiality. In J. Selzer & S. Crowley (Eds.), *Rhetorical bodies* (pp. 16–57). Madison: University of Wisconsin Press.

Blakesley, D., & Rickert, T. (2004). From nodes to nets: Our emerging culture of complex interactive networks. *JAC, 24*(4), 821–830.

Boyle, C. (2016). Writing and rhetoric and/as posthuman practice. *College English, 78*(6), 532–554.

Braidotti, R. (2013). *The posthuman.* Cambridge, UK: Polity Press.

Breu, C. (2014). *Insistence of the material: Literature in the age of biopolitics.* Minneapolis: University of Minnesota Press.

Brooke, C. G. (2009). *Lingua fracta: Towards a rhetoric of new media.* New York, NY: Hampton Press.

Brown, J. J., Jr., & Rivers, N. A. (2013). Composing the carpenter's workshop. *O-Zone, 1*(1), 27–36.

Butler, J. (1993). *Bodies that matter: On the discursive limits of "sex."* New York, NY: Routledge.

Campbell, N. D. (2009). Reconstructing science and technology studies: Views from feminist standpoint theory. *Frontiers, 30*(1), 1–29.

Ceccarelli, L. (2001). *Shaping science with rhetoric: The cases of Dobzhansky, Schrödinger, and Wilson.* Chicago, IL: University of Chicago Press.

Chen, M. Y. (2012). *Animacies: Biopolitics, racial mattering, and queer affect.* Durham, NC: Duke University Press.

Cisneros, J. (2008, January). *(Re)making the immigrant body: Rhetoric, materiality, and social protest in "La Gran Marcha" of March 25, 2006.* Paper presented at the meeting of the National Communication Association, San Diego, CA. Accessed 9 June 2016. Retrieved from http://citation.allacademic.com/meta/p_mla_apa_research_citation/2/6/0/3/7/pages260372/p260372-26.php

Condit, C. (1996). How bad science stays that way: Brain sex, demarcation, and the status of truth in the rhetoric of science. *Rhetoric Society Quarterly, 26*(4), 83–109.

Condit, C. M. (1994). Hegemony in a mass-mediated society: Concordance about reproductive technologies. *Critical Studies in Mass Communication, 11*(3), 205–230.

Condit, C. M. (1999). *The meanings of the gene: Public debates about human heredity.* Madison: University of Wisconsin Press.

Condit, C. M. (2004). *How should we study the symbolizing animal?* Boston, MA: Pearson Education.

Condit, C. M., Lynch, J., & Winderman, E. (2012). Recent rhetorical studies in public understanding of science: Multiple purposes and strengths. *Public Understanding of Science, 21*(4), 386–400.

Coole, D., & Frost, S. (Eds.). (2010). *New materialisms: Ontology, agency, and politics.* Durham, NC: Duke University Press.

Cooper, M. M. (2011). Rhetorical agency as emergent and enacted. *College Composition and Communication, 62*(3), 420–449.

Dingo, R. (2012). *Networking arguments: Rhetoric, transnational feminism, and public policy writing.* Pittsburgh, PA: University of Pittsburgh Press.

Dobrin, S. I. (Ed.). (2015). *Writing posthumanism, posthuman writing.* Anderson, SC: Parlor Press.

Dolphijn, R., & van der Tuin, I. (2012). *New materialism: Interviews and cartographies.* Ann Arbor, MI: Open Humanities Press.

Edbauer, J. (2005). Unframing models of distribution: From rhetorical situation to rhetorical ecologies. *Rhetoric Society Quarterly, 35*(4), 5–24.

Fahnestock, J. (1996). Series reasoning in scientific argument: *Incrementum* and *gradatio* and the case of Darwin. *Rhetoric Society Quarterly, 26*(4), 13–40.

Fahnestock, J. (2013). Promoting the discipline: Rhetorical studies of science, technology, and medicine. *Poroi, 9*(1). Accessed 22 July 2015. Retrieved from http://ir.uiowa.edu/poroi/vol9/iss1/6/

Fausto-Sterling, A. (2000). *Sexing the body: Gender politics and the construction of sexuality.* New York, NY: Basic Books.

Fleckenstein, K. S. (1999). Writing bodies: Somatic mind in composition studies. *College English, 61*(3), 281–306.

Flores, L. A. (1996). Creating discursive space through a rhetoric of difference: Chicana feminists craft a homeland. *Quarterly Journal of Speech, 82*(2), 142–156.

Foucault, M. (1977). *Discipline and punish: The birth of the prison.* (A. Sheridan, Trans.). London, UK: Allen Lane.

Foucault, M. (1990). *The history of sexuality: Vol. 1. An introduction.* (R. Hurley, Trans.). New York, NY: Vintage Books.

Frost, S. (2011). The implications of the new materialisms for feminist epistemology. In H. E. Grasswick (Ed.), *Feminist epistemology and philosophy of science: Power in knowledge* (pp. 69–82). London, UK: Springer.

Grabill, J. (2014). The work of rhetoric in the common places: An essay on rhetorical methodology. *JAC, 34*(1–2), 247–267.

Greene, R. W. (1998). Another materialist rhetoric. *Critical Studies in Mass Communication, 15*(1), 21–40.

Gries, L. E. (2015). *Still life with rhetoric: A new materialist approach for visual rhetorics.* Boulder, CO: Utah State University Press.

Gronnvoll, M. (2013). Material rhetorics meet material feminisms. *Quarterly Journal of Speech, 99*(1), 98–113.

Gross, A. G. (1996). *The rhetoric of science.* Cambridge, MA: Harvard University Press.

Grosz, E. (2008). Darwin and feminism: Preliminary investigations for a possible alliance. In S. Alaimo & S. Hekman (Eds.), *Material feminisms* (pp. 23–51). Bloomington: Indiana University Press.

Grosz, E. (2010). Feminism, materialism, and freedom. In D. Coole & S. Frost (Eds.), *New materialisms: Ontology, agency, and politics* (pp. 139–157). Durham, NC: Duke University Press.

Hallenbeck, S. (2012). Toward a posthuman perspective: Feminist rhetorical methodologies and everyday practices. *Advances in the History of Rhetoric, 15*(1), 9–27.

Hallenbeck, S. (2015). *Claiming the bicycle: Women, rhetoric, and technology in nineteenth-century America.* Carbondale: Southern Illinois University Press.

Halloran, S. M. (1984). The birth of molecular biology: An essay in the rhetorical criticism of scientific discourse. *Rhetoric Review, 3*(1), 70–83.

Haraway, D. (1988). Situated knowledges: The science question in feminism and the privilege of partial perspective. *Feminist Studies, 14*(3), 575–599.

Haraway, D. J. (1991). *Simians, cyborgs, and women: The reinvention of nature.* New York, NY: Routledge.

Harding, S. (1986). *The science question in feminism.* Ithaca, NY: Cornell University Press.

Harding, S. (1991). *Whose science? Whose knowledge? Thinking from women's lives.* Ithaca, NY: Cornell University Press.

Harding, S. (1992). Rethinking standpoint epistemology: "What is strong objectivity"? In L. Alcoff & E. Potter (Eds.), *Feminist Epistemologies* (pp. 49–82). New York, NY: Routledge.

Harding, S. (Ed.). (2011). *The postcolonial science and technology studies reader.* Durham, NC: Duke University Press.

Hawk, B. (2011). Reassembling postprocess: Toward a posthuman theory of public rhetoric. In S. I. Dobrin, J. A. Rice, & M. Vastola (Eds.), *Beyond postprocess* (pp. 75–93). Boulder, CO: Utah State University Press.

Hawk, B., Lindgren, C., & Mara, A. (2015). Utopian laptop initiatives: From technological deism to object-oriented rhetoric. In S. I. Dobrin (Ed.), *Writing posthumanism, posthuman writing* (pp. 192–213). Anderson, SC: Parlor Press.

Hayles, N. K. (1999). *How we became posthuman: Virtual bodies in cybernetics, literature, and informatics.* Chicago, IL: University of Chicago Press.

Hekman, S. (2010). *The material of knowledge: Feminist disclosures.* Bloomington: Indiana University Press.

Helmers, M. (2006). Objects, memory, and narrative: New notes toward materialist rhetoric. In K. Ronald & J. Ritchie (Eds.), *Teaching rhetorica: Theory, pedagogy, practice* (pp. 114–130). Portsmouth, NH: Boynton/Cook.

Herndl, C. G., & Cutlip, L. L. (2013). "How can we act?" A praxiographical program for the rhetoric of technology, science, and medicine. *Poroi, 9*(1).

Accessed 22 July 2015. Retrieved from http://ir.uiowa.edu/cgi/viewcontent.cgi?article=1163&context=poroi

Hesford, W. S. (1999). Reading rape stories: Material rhetoric and the trauma of representation. *College English, 62*(2), 192–221.

Holmes, S. (2013). *Actants, agents, and assemblages: Delivery and writing in an age of new media* (Doctoral dissertation). Retrieved from ProQuest Dissertations and Theses Global. (Document No. 1413296672)

Jung, J. (2014). Systems rhetoric: A dynamic coupling of explanation and description. *Enculturation, 17*(1). Accessed 26 July 2015. Retrieved from http://www.enculturation.net/systems-rhetoric

Keller, E. F. (1985). *Reflections on gender and science.* New Haven, CT: Yale University Press.

Keränen, L. (2013). Conspectus: Inventing futures for the rhetoric of science, technology, and medicine. *Poroi: An Interdisciplinary Journal of Rhetorical Analysis and Invention, 9*(1). Accessed 22 July 2015. Retrieved from http://ir.uiowa.edu/poroi/vol9/iss1/1/

Kessler, M. M. (2016). Wearing an ostomy pouch and becoming an ostomate: A kairological approach to wearability. *Rhetoric Society Quarterly, 46*(3), 236–250.

Longino, H. E. (1990). *Science as social knowledge.* Princeton, NJ: Princeton University Press.

Lykke, N. (2008). Feminist cultural studies of technoscience: Portrait of an implosion. In A. Smelik & N. Lykke (Eds.), *Bits of life: Feminism at the intersections of media, bioscience, and technology* (pp. 3–15). Seattle: University of Washington Press.

Lynch, P., & Rivers, N. (Eds.). (2015). *Thinking with Bruno Latour in rhetoric and composition.* Carbondale: Southern Illinois University Press.

Marback, R. (2008). Unclenching the fist: Embodying rhetoric and giving objects their due. *Rhetoric Society Quarterly, 38*(1), 46–65.

Mays, C. (2015). From "flows" to "excess": On stability, stubbornness, and blockage in rhetorical ecologies. *Enculturation, 19.* Accessed 21 December 2016. Retrieved from http://enculturation.net/from-flows-to-excess

McGee, M. C. (1982). A materialist's conception of rhetoric. In R. McKerrow (Ed.), *Explorations in rhetoric: Studies in honor of Douglas Ehringer* (pp. 23–48). Glenview, IL: Scott, Foresman.

Mol, A. (2002). *The body multiple: Ontology in medical practice.* Durham, NC: Duke University Press.

Myers, G. (1990). *Writing biology: Texts in the social construction of scientific knowledge.* Madison: University of Wisconsin Press.

Nicotra, J. (2016). Assemblage rhetorics: Creating new frameworks for rhetorical action. In S. Barnett & C. Boyle (Eds.), *Rhetoric, through everyday things* (pp. 185–196). Tuscaloosa: University of Alabama Press.

Ortoleva, M. (2013). Let's not forget ecological literacy. *LiSC, 1*(2), 66–73.

Oudshoorn, N., Brouns, M., & van Oost, E. (2005). Diversity and distributed agency in the design and use of medical video-communication technologies. In H. Harbers (Ed.), *Inside the politics of technology: Agency and normativity in the co-production of technology and society* (pp. 85–105). Amsterdam, Netherlands: Amsterdam University Press.

Prasad, A. (2014). *Imperial technoscience: Transnational histories of MRI in the United States, Britain, and India.* Cambridge: Massachusetts Institute of Technology Press.

Rice, J. (2006). Networks and new media. *College English, 69*(2), 127–133.

Richardson, S. S. (2010). Feminist philosophy of science: History, contributions, and challenges. *Synthese, 177*(3), 337–362.

Rickert, T. (2013). *Ambient rhetoric: The attunements of rhetorical being.* Pittsburgh, PA: University of Pittsburgh Press.

Ridolfo, J., & Hart-Davidson, W. (Eds.). (2015). *Rhetoric and the digital humanities.* Chicago, IL: University of Chicago Press.

Ríos, G. R. (2015). Cultivating land-based literacies and rhetorics. *LiCS, 3*(1), 60–70.

Rivers, N. A., & Weber, R. P. (2011). Ecological, pedagogical, public rhetoric. *College Composition and Communication, 63*(2), 187–218.

Roberts, D. (1997). *Killing the black body: Race, reproduction, and the meaning of liberty.* New York, NY: Pantheon.

Scott, J. B. (2003). *Risky rhetoric: AIDS and the cultural practices of HIV testing.* Carbondale: Southern Illinois University Press.

Selzer, J., & Crowley, S. (Eds.). (1999). *Rhetorical bodies.* Madison: University of Wisconsin Press.

Sharon, T. (2014). *Human nature in an age of biotechnology: The case for mediated posthumanism.* New York, NY: Springer.

Shepley, N. (2013). Rhetorical-ecological links in composition history. *Enculturation, 15.* Accessed 21 December 2016. Retrieved from http://enculturation.net/rhetorical-ecological-links

Simonsen, K. (2012). In quest of a new humanism: Embodiment, experience, and phenomenology as critical geography. *Progress in Human Geography, 37*(1), 10–26.

Smelik, A., & Lykke, N. (Eds.). (2008). *Bits of life: Feminism at the intersections of media, bioscience, and technology.* Seattle: University of Washington Press.

Spurr, D. (1993). *The rhetoric of empire: Colonial discourse in journalism, travel writing, and imperial administration.* Durham, NC: Duke University Press.

Subramaniam, B. (2009). Moored metamorphoses: A retrospective essay on feminist science studies. *Signs, 34*(4), 951–980.

Takeshita, C. (2011). *The global biopolitics of the IUD: How science constructs contraceptive users and women's bodies.* Cambridge: Massachusetts Institute of Technology Press.

Teston, C. (2016). Rhetoric, precarity, and mhealth technologies. *Rhetoric Society Quarterly, 46*(3), 251–268.

Thornton, D. J. (2011). *Brain culture: Neuroscience and popular media.* Piscataway, NJ: Rutgers University Press.

Trader, K. S. (2013). Assuming differently: Posthumanism, enthymeme, and the possibility of change. *JAC, 33*(1/2), 201–231.

Tuana, N. (Ed.). (1989). *Feminism and science.* Bloomington: Indiana University Press.

Wajcman, J. (2010). Feminist theories of technology. *Cambridge Journal of Economics, 34*(1), 143–152.

Washick, B., & Wingrove, E. (2015). Ontologized agency and political critique. *Contemporary Political Theory, 14*(1), 63–79.

Weheliye, A. G. (2014). *Habeas viscus: Racializing assemblages, biopolitics, and black feminist theories of the human.* Durham, NC: Duke University Press.

Whelan, E. (2001). Politics by other means: Feminism and mainstream science studies. *Canadian Journal of Sociology, 26*(4), 535–581.

White-Farnham, J. (2013). Changing perceptions, changing conditions: The material rhetoric of the Red Hat Society. *Rhetoric Review, 32*(4), 473–489.

Wilson, J. C., & Lewiecki-Wilson, C. (Eds.). (2001). *Embodied rhetorics: Disability in language and culture.* Carbondale: Southern Illinois University Press.

Wyer, M., Barbercheck, M., Cookmeyer, D., Öztürk, H. Ö., & Wayne, M. (Eds.). (2014). *Women, science, and technology: A reader in feminist science studies* (3rd ed.). New York, NY: Routledge.

1.

Of Complexity and Caution

Feminism, Object-Oriented Ontology, and the Practices of Scholarly Work

Kyle P. Vealey and Alex Layne

> The term "objects" is not opposed to "subjects," so it is not such a bad fate to be an object.
> —Graham Harman, *Bells and Whistles*

> Let me be clear: we need not discount human beings to adopt an object-oriented position—after all, we ourselves are of the world as much as musket buckshot and gypsum and space shuttles.
> —Ian Bogost, *Alien Phenomenology, or What It's Like to Be a Thing*

> In an atmosphere of expansion and bandwagons, there is little room for complexity and caution.
> —Susan Leigh Star, *Ecologies of Knowledge*

Object-oriented ontology (OOO), an emerging movement associated with Graham Harman, Levi Bryant, Ian Bogost, and Timothy Morton, has gained considerable traction in the past few years. Overall, OOO aims to counteract a philosophically inherited understanding of objects as accessible only through their relations with humans. Building from Martin Heidegger's tool-analysis in *Being and Time* (2008), OOO sees objects not only relating *to* and *for us* but also agentially relating to one another equally and vibrantly. As Bogost articulates in *Alien Phenomenology, or What It's Like to Be a Thing* (2012), "OOO puts *things* at the center of being . . . [and] contends that nothing has special status, but that everything exists equally—plumbers, cotton, bonobos, DVD players, and sandstone" (p. 6). This democratizing of ontology allows for a more robust philosophical investigation not only between humans and nonhumans but also between nonhuman and nonhuman that fundamentally resists reducing matter to mere aggregates of physical forces. Furthermore, proponents of OOO remain steadfast in challenging simplistic container metaphors—such as "nature," "world," "universe"—that, in one fell swoop, gather up the multiplicity and diversity of existing objects. Following and adapting models proposed in Bruno Latour's "Irreductions" from *The Pasteurization of France* (1993) and Manuel DeLanda's *Intensive Science and Virtual Philosophy* (2002), OOO attends to the unique and individuated object, thus placing all entities, from frogs, nootropics, and nuclear warheads to breast pumps, decanters, and HIV, on equal ontological footing. This egalitarian move, what DeLanda calls a "flat ontology," demonstrates the inclusive and ontologically indiscriminate orientation that drives much, if not all, of OOO's work (p. 47).

On numerous occasions, in both print and online, the four core members of the movement have described the interdisciplinary tone, expansive application, and deeply inclusive nature of OOO. Bryant (2010) comments that

> OOO is among the most open philosophical movements that's ever existed. On the one hand, OOO has generated a large inter-disciplinary interest from people both inside and outside the academy. Not only has OOO drawn interest from rhetoricians, anthropologists, media theorists, literary theorists, biologists, and even a handful of physicists, it has also drawn the interest of artists, activists, feminists, and so on.

We want to highlight two points in Bryant's description, one we find productive and one we caution against. OOO's push for interdisciplinary engagement with rhetoric studies is admirable, particularly in light of the well-known (and perhaps still too well maintained) division between rhetoric and philosophy. And while rhetorical theory throughout the years has exhibited significant interest in materiality (Haas, 1996; Horner, 2000; Wysocki, 2004), embodiment (DeVoss, Cushman, & Grabill, 2005; Rice, 2008), and the complex material contexts of rhetorical work (Cooper, 1986; Edbauer, 2005; Syverson, 1999; Reid, 2013; Rickert, 2013), recent scholarship has sought to describe and build on the intersections of rhetorical theory and OOO. For instance, Scot Barnett argues in his 2010 review of Harman's *Tool-Being* (2002) and *Guerrilla Metaphysics* (2005), in which he coined the phrase "object-oriented rhetoric," that

> the idea of an object-oriented rhetoric compels us to re/consider the very nature of rhetoric itself and to think carefully through what the implications

our missing masses suggest about rhetoric as both a human art and an ontological condition potentially operable alongside human beings in the world's vast and inexhaustible carpentry of things.

While we recognize the value of this particular interdisciplinary engagement and the abundant space it opens up for thinking and talking about the inborn rhetoricity of objects and environments, we likewise notice a rhetorical positioning in Bryant's description that arguably works as a representative anecdote for OOO as a whole. In an attempt to drum up OOO's extensive range of application, Bryant makes a somewhat unnecessary division between academic work, on the one hand, and activist work, on the other, resulting in a boundary being drawn between scholarly work in the academy and the rich history of feminism and feminist thought. The division is odd and—whether intentional or not—warrants further consideration of OOO's consistent lack of engagement with feminist thought as well as the ways in which it constructs its own philosophical history (one primarily rooted in philosophers such as Heidegger and Latour) through an exclusion of feminist scholars. Moreover, Bryant's division calls attention to the fact that ontology is not merely a topic of scholarly inquiry—rather, ontologies are always built and supported through practices of scholarly work. In the context of feminist rhetorical science studies (FRSS), such practices are not socially and politically neutral but are always involved in the construction of boundaries between what matters and what does not. As Karen Barad (2007) describes, such discursive practices "[constrain] and [enable] what can be said . . . [and] define what counts as meaningful statements" in local contexts (p. 146).

In light of the problems we see in OOO's relationship to feminist thought, we advocate for a shift from imagining ontology as strictly a topic of inquiry to foregrounding the ways ontologies are built and sustained through mundane and often invisible practices of scholarly work. Scholarly work, for us, encompasses the full range of practices involved in the production of research and scholarship. Practices of scholarly work, then, are highly local and situated, idiosyncratic, amorphous, messy, and iterative. They are, in other words, the *doing* of scholarship. Most importantly, scholarly work defines and continually redefines its borders with its citation practices, mapping the intellectual camps, theories, and authors that compose the landscape on which authors situate their work. Through citational practices, authors tether themselves (or express their indebtedness) to others and ideas that have come before and, in some way, shaped their own thinking and scholarship. We further elaborate this point later in the chapter.

Citation practices are a particularly salient issue for FRSS and feminist rhetorics more broadly. Reflecting on how their own location at the intersection of feminisms and rhetorics impacts their scholarly practices, Jacqueline Jones Royster and Gesa E. Kirsch (2012) observe:

> Our sense of researchers and scholars not having the authority, privilege, and entitlement to write or write over the presence of others is pushing us rather persistently to operationalize this view in our actual ways of thinking, being, and doing as professionals in the field. (pp. 144–145)

They specifically call on feminist rhetorics scholars to be "deliberate about developing and sustaining throughout the analytical process a more conscious and explicit habit of thinking about our work as part of, rather than disconnected from, other rhetorical enterprises around the world" (p. 145). One habit of mind they propose involves considering the *social circulation* of scholarship, which they understand as an "analytical effort . . . to pay attention to the ways that ideas travel in order for us to become more consciously aware of patterns of intellectual and social engagement" (p. 138). Such a habit of mind demands that scholars continually reflect on the social and political impact of their work in the hopes of thinking "beyond the concrete in envisioning alternative possibilities in order that we might actually work, often collaboratively, toward enacting a better future" (p. 145).

Building on Royster and Kirsch's portrait of a new methodological landscape, this chapter aims to articulate a feminist rhetorical methodology that understands scholarly work as a series of practices that constitute, rather than reflect, ontologies. Specifically, we propose a way of remaining steadfast in understanding FRSS scholarship as (*1*) an object composed in and through its relations to past scholarly work; and (*2*) an enactment of particular configurations of the world that have social and political effects on others. We contextualize the need for this methodology in a critique of OOO's undocumented relationship to feminist thought. In doing so, we set out to explore rhetorical ways of calling attention to the social and political dimensions of scholarly work. We suggest, in other words, that productive and equitable engagement in discussions of the material world calls for a sustainable approach to ontology that keeps the social and political implications of our scholarly work at the foreground of our minds. Our methodology is, in many ways, indebted to Barad's (2007) notion that ethics is always entangled with ontology and epistemology, thus characterizing ethics as not solely the determination of correct or right responses to specific situations but rather as an ongoing concern for "taking account of entanglements" that configure and reconfigure the world (p. 384).

Our chapter is divided into three parts. First, we examine discussions of OOO as an emerging school of realist thought. We focus on criticisms lodged against OOO from a number of feminist scholars, such as Judith Halberstam, Rosi Braidotti, and Jussi Parikka, which call attention to OOO's lack of engagement with feminist theory. We then contextualize these criticisms in a broader discussion of the mundane and often invisible practices of scholarly work. Second, we draw connections between OOO's lack of engagement with feminist thought and the ontological commitments espoused in their scholarship. In doing so, we look at specific pieces of feminist scholarship that grapple with the ontology of objects and their relations. While this section does not do justice to the myriad of existing feminist scholarship, we choose several pieces that we believe to be missed opportunities for OOO to engage productively with feminist thought. Third, we conclude by developing a feminist rhetorical methodology for approaching scholarly work as a complex series of ontological enactments. Our methodology is grounded in a habit of mind that foregrounds two key questions: what realities are produced through scholarly work, and whom do those realities directly or indirectly serve? This habit of mind is a reflexive consideration of the *effects* of scholarly work, or what we call the "rhetorical reverberations" scholarship has in the world. We propose this methodology not as a prescriptive notion that needs to be followed but more as a call to action, a call that opens up space for more complexity, caution, and care in the production of feminist rhetorical science scholarship.

Object-Oriented Ontology and the Politics of Scholarly Practices

While feminist thought has sustained an engagement with materiality for over thirty years, the recent resurgence of interest in things and objects has taken roughly three forms. First, literary scholars in the 1990s began to reconfigure their interests in objects, shifting away from exploring their hidden and cultural purposes toward determining their power as circulators of both human and nonhuman meaning. This interest culminated in Bill Brown's "thing theory," notably articulated in a 2001 special issue of *Critical Inquiry*, where he suggests that our increasingly digitized world has necessitated a critical view of physical, tangible objects (p. 16). Things, for Brown, pose problems, continually unsettling themselves as silent surveyors of human activities; through their "specific unspecificity," things both escape and exceed our attempts to reduce them to a simple subject/object split while continually appearing with a "suddenness" that expresses their "presence and power"

(p. 3). Things are, in other words, evocative of complex meanings that we cannot simply ignore.

Second, "new materialism," a term used respectively by Braidotti and DeLanda in the 1990s, describes a diverse body of theoretical, political, and scientific work that challenges traditionally dichotomous understandings of nature/culture, mind/body, and human/nonhuman (Dolphijn & van der Tuin, 2012, p. 48, 93). At the center of new materialist thought is the work of Donna Haraway, N. Katherine Hayles, and Nancy Tuana, among many others. Diana Coole and Samantha Frost, in their collection *New Materialisms: Ontology, Agency, and Politics* (2010), suggest that new materialism differentiates itself from "the exhaustion of once popular materialist approaches, such as existential phenomenology or structural Marxism" (p. 3), shifting the focus to the "forces, energies, and intensities (rather than substances)" that make up our very material lives (p. 13).

Finally, emerging from the field of philosophy, OOO seeks to reinstate a sense of "thingness" to the many objects that litter and make up the world, particularly in a way that is not exclusively tethered to human modes of access, use, or meaning. As Harman indicates in *Tool-Being* (2002), the tool-use section in *Being and Time* metaphorically "gives birth to an ontology of *things themselves*" (p. 1).

OOO, while emerging during this exciting and much talked about explosion of feminist scholarship on the material, seems to firmly doubt any influence of such work. Certainly this work has affected many other fields and disciplinary ways of thinking, and it undoubtedly has had an impact, however unspoken, on the emergence of an object-oriented way of thinking. It is not that OOO scholars find themselves opposed to how feminist and new materialist scholars theorize objects, but rather that they often position themselves as emerging from a scholarly group that does not engage with work labeled "feminist." The feminists who have spent years doing similar work ironically have little ontology. To begin to illustrate the unspoken connections between feminist work and OOO, the next section explores OOO's emergence, attempting to parse the nuances of OOO as a philosophy, a collection of scholars, and an object itself that has material consequences. Of course, any positioning of a scholarly movement as stable, as having distinct predecessors, or as having inert relations is always more complicated than can be seen on the surface. That said, OOO's visible patterns of persistently ignoring the influence of feminist scholarship create troubling inconsistencies within its own ontology.

Tenets of Object-Oriented Ontology

OOO emerged as an offshoot of a broader school of thought known as speculative realism (SR). As Harman notes, SR "has always been a loose umbrella term for four markedly different positions: my own object-oriented philosophy, Ray Brassier's eliminative nihilism, Iain Hamilton Grant's cyber-vitalism, and Quentin Meillassoux's speculative materialism" (2010, *Towards*, p. 1). Hence, we can say that SR and OOO assume a decidedly firm stance against what Meillassoux calls correlationist thought—that is, a view of the world existing only for humans, with humans existing a priori. Harman (2009) articulates:

> The correlationist holds that we cannot think of humans without world, nor world without humans, but only of a primal correlation or rapport between the two.... Correlationism is neither materialism nor absolute idealism, since it glues human and world together from the start, giving preference to neither. (pp. 122–123)

An anticorrelationist stance, in contrast, largely decenters (but does not entirely displace) the human from philosophical investigations and abandons the belief that human access forever conditions how we think, talk, and move through the world.

While SR remains a broad category encompassing an array of diverse philosophical perspectives, OOO has developed as a tight-knit circle of philosophers who are as quick as they are prolific in their scholarly work. Their scholarship enacts divisions and boundary-making practices that influence the way OOO exists in the world as well as its impact on other philosophies and scholars. To be an object-oriented ontologist, according to Harman (2013), "what you need to do is hold that individual entities of various different scales are the ultimate stuff of the cosmos" (p. 6). While this basic principle unites OOO as a philosophical movement, Harman differentiates his own brand of object-oriented thought through an additional premise that such objects "are never exhausted by any of their relations or even by their sum of all possible relations" (p. 7). As Harman articulates it, OOO charts a middle route through two predominant approaches to understanding the material world: undermining and overmining. Undermining, Harman describes, is a kind of "dogmatic reductionism, failing to see that mid-sized levels of the world can have their own autonomy, are often partially independent of their tinier constituent pieces, and can affect their own pieces or even generate new ones" (p. 35). Overmining, on the other hand, is a kind of materialism "generally of a cultural rather than physical sort" that sees "objects only . . .

in some social or linguistic context. They are purely relational" (p. 35). At the core of Harman's OOO is a focus on the ways objects withdraw from their relations. Objects are, in other words, autonomous entities that are not reducible to surrounding objects or as finite substances.

Levi Bryant (2011, *Democracy*) instead offers what he calls an "onticology," a kind of flat ontology that he summarizes in four theses. First, a flat ontology resists all attempts to privilege one particular, transcendent object over and above the rest (p. 245). That is to say, there does not exist one particular object that transcends and rules them all. Second, Bryant contends that there is no such thing as *the* world or *the* universe (p. 246). The emphasis in this second tenet on the article *the* designates one particular entity that acts as a container for all that exists. Third, a flat ontology resists the privileging of one kind of relation over others. For example, there is no ontological difference in kind between a subject-object relation and an object-object relation (p. 246). The important point in this third thesis is that a flat ontology contests the notion that human subjects have to be present in any and all object-object relations. To reject this claim is to accept correlationism, which Bryant and other proponents of OOO are committed to challenging. Fourth, a flat ontology abides by an understanding that all entities exist on equal ontological footing. That is, objects such as strawberries, staph infections, locomotives, and genomes are all equally existing objects that do not ontologically differ in degree or kind from one another. Bryant's intention in developing a flat ontology is not to debunk or usurp prevailing social, political, and ethical notions in continental philosophy. Rather, he describes the ideal flat ontology as being a synthesis of two prevailing cultures in critical theory and philosophy: the first focusing on "lived experience, text, discourse, signifiers, signs and representation, and meaning" and the second considering a wide range of "nonhuman actors or objects" (p. 247). Importantly, Bryant states that his democracy of objects is not offered as a "*political* thesis to the effect that all objects ought to be treated equally" but as a way to widen the spectrum of actors considered in important social, cultural, and political discussions (p. 19).

Following Bryant's second premise of a flat ontology, Timothy Morton's (2012) contribution to OOO scholarship proposes that we let go of the notion that *the* world, or Nature, exists and instead embrace what he calls "the ecological thought" that breaks us from fictions that would place humans as uniquely seated over and above the material world (p. 3). Thinking the ecological thought means "thinking of interconnectedness.... It's a practice and a process of becoming fully aware of how human beings are connected with other beings—animal, vegetable, or mineral" (p. 7). In his more recent *Hyperobjects: Philosophy and Ecology after the End of the World* (2013), Morton pushes the notion of the ecological thought further by portraying landmark events in ecological and geopolitical history—moments ranging from the early development of Little Boy and Fat Man and their use in 1945 to ongoing efforts in fracking for oil taking place across the globe—as direct, philosophical legacies of correlationist thought that limits thinking solely to correlations between human and world (p. 10). For too long, Morton suggests, we have accepted an ontology that has left us stuck within ourselves. However, with the emergence of what he calls "hyperobjects"—nonhuman entities that are "massively distributed in time and space relative to humans," such as global warming, black holes, and nonbiodegradable plastics—we have found ourselves in need of a new way to think and talk about the material world as an active agent (p. 1).

Like Morton's development of hyperobjects, Ian Bogost also builds on the basic principles of OOO in his *Alien Phenomenology, or What It's Like to Be a Thing* (2012) by asking how we have come to live in an "era in which 'things' means ideas so often, and stuff so seldom" (p. 3). Bogost explores this question by chronicling the emergence of OOO in response to Kantian philosophy and other correlationist perspectives that see the world as always filtered through human access. Situating his project within an OOO framework, Bogost proposes a number of speculative and imaginative possibilities that an object-oriented realism opens up for philosophers. For example, he points to carpentry as capable of becoming a "rigorous kind of philosophical creativity," particularly because "it rejects the correlationist agenda by definition, refusing to address only the human reader's ability to pass eyeballs over words and intellect over notions they contain" (pp. 92–93). *Alien Phenomenology* offers numerous ways that OOO can be put into practice in order to speculate, as Bogost's subtitle suggests, on what it's like to be a thing.

Feminist Critiques of Object-Oriented Ontology

OOO's prolific outpouring across various blogs and other forms of social media as well as in print and electronic journals has allowed scholars from a variety of fields to engage in sustained conversations about the philosophy's basic tenets. These public exchanges have also provided ample space for critique. Common to a number of these critiques coming from prominent feminist thinkers is the claim that OOO has not adequately engaged with aspects of materiality that have been at the center of feminist thought since the 1970s. Judith Halberstam (2012), for one, notes that the theories that count and that get counted in OOO and SR tend to be masculinist most of the time and tend to cluster around enlightenment and post-structuralist theory or a particular, continental stripe: Hegel, Heidegger, Derrida, Žižek, Lacan, with a Butler or Braidotti thrown in for good measure but nary a mention of race, class or postcolonial thinking.

And although she is sometimes mentioned, even Braidotti believes her work has gone unaddressed in conversations regarding SR and OOO. In a 2014 interview with Timotheus Vermeulen, Braidotti critiques both movements because they have not fully engaged with well-established discussions of materiality: "Media and science fiction scholars—like Jussi Parikka now, or Donna Haraway before him—have been theorizing objects along these lines for years." Braidotti's critique specifically calls attention to practices enacted in OOO and SR scholarship:

> I am surprised, sometimes even shocked, that their discussions and bibliographies make little mention of these debates. . . . 1970s feminism: What is that? It's a planet, it's a galaxy. It includes De Beauvoir, Irigaray and Deleuzian studies. The disrespect, the competitiveness: that's bad scholarship. This really needs to be said because it makes the conversation extremely difficult. I've read the stuff; I do my duty. I doubt they have ever read anything I wrote but, if they have, it doesn't show.

Braidotti's comment here stresses that the problem is less about a lack of recognition for recognition's sake and more about creating politically fraught obstacles that prohibit conversations between OOO and feminism. In other words, it's not about ego but about creating ethical conversations and possibilities for reciprocity in research.

Parikka, as early as 2011, attempted to initiate one such conversation by asking a series of blog-style "Object-Oriented-Questions," one of which reads:

> Some people are enthusiastic because object oriented philosophy seems at last to offer a philosophical way of treating the non-human (animals, technology, etc.) on an equal footing to the human. Agencies are extended to a whole lot of entities. But such claims, whether intentionally or not, forget that there is a whole long history of such thought; the most often forgotten is the radical feminist materialism of figures such as Rosi Braidotti and Elizabeth Grosz; this goes nowadays often by the name of new materialism.

Parikka's post received a number of replies from OOO scholars. Harman (2011), for instance, responded to the concern that OOO has ignored a long history of feminist thought by avoiding the criticism altogether and instead observing that Parikka's

> blog comments don't cite any OOO writings at all, though the questions he raises are already openly addressed in most of them. A more productive means of engagement, I think, would be actually to read the OOO arguments in favor of an object-oriented approach and then say why one finds them lacking.

Bryant's response acknowledged the productive similarities between feminist new materialism and OOO, noting that

> this is one of the major reasons that we invited Elizabeth Grosz to be on O-Zone's editorial board. We recognized that there are fruitful points of overlap among these different intellectual trajectories and that they should be brought into dialogue with one another. (2011, "Some Responses")

In a similar fashion, Bogost (2011) responded by saying:

> We all have our influences. I read this question as Parikka's declaration that he also likes Braidotti and Grosz on these subjects, and that's great. The more the merrier. While this is not necessarily the place to go into detail, I think there are both similarities and differences between OOO and the new feminist materialism, enough of both to make for productive conversation.

While each of the responders was careful to address Parikka's questions regarding the principles that make up an OOO, their replies to his criticism—that OOO has ignored decades of feminist work on materiality—amount to little more than a passing acknowledgment that there are indeed productive connections to be made, rather than a serious call for engagement and change. A brief look at the bibliographies of major OOO books confirms the critiques lodged by Halberstam, Braidotti, and Parikka. For instance, Harman's *Bells and Whistles* (2013) cites 2 women out of 242 total citations. Bryant's *Democracy of Objects* (2011) cites 8 works by women out of 146 total references, while Bogost's *Alien Phenomenology* (2012) cites 8 women out of 111 total citations.[1]

One might object here and say that critiquing mundane scholarly practices such as citations is, at best, getting caught up in trivial details or, at worst, bordering on an ad hominem argument. But because citations and citation practices often define the borders of knowledge, they are an important part of how scholarly work situates itself in terms of past, present, and future inquiries. Indeed, our focus on particular citational practices, in many ways, resonates with postmodernism's sustained engagement with citationality and the ontological weight citations carry. For instance, the notion of citationality plays a large role in what Jacques Derrida (1988) understood as the graphematic structure of language, which sees citationality not as "the same sort as in a theatrical play, a philosophical reference, or the recitation of a poem" but rather as what constitutes the fundamental characteristics of language (p. 18, 23). Derrida's notion of citationality is later picked up and employed by Judith Butler (1990, 1993) in her argument for shifting how we understand both sex and gender from preestablished categories of fact to discursive constructions shaped and reshaped by cultural inscription. Butler's *Bodies That Matter* (1993), for instance, foregrounds the ways sex is discursively constructed through an array of cultural practices that are, at their core, constituted by their citationality, or iterability, over time. Sex, in other words, is not a prelinguistic fact; rather, it is more appropriately conceived of as a socially inscribed construct that, over time, materializes through repetitive and citational adherence to an established and oppressive regulatory schema (p. 1). However, Butler's argument is ultimately rooted in the idea that such citationality (at the core of both linguistic signs and cultural practices) can also be used in subversive ways. That is, Butler sees citationality as a source of strategies for achieving political agency by continually revealing how everyday citational practices constitute how we understand, engage, and embody sex and gender in day-to-day life.

Critiques of citational practices also have a long history in feminist work in science and technology studies, where citations are seen as defining the borders of disciplinary knowledge. Sara Ahmed (2013), for one, describes citation practices as "*screening techniques*," ways to make "certain bodies take up spaces by screening out the existence of others. If you are screened out (by virtue of the body you have) then you simply do not even appear or register to others." Responding to those who regard this form of critique as "very 1980s," Ahmed argues that we still need

> feminist and anti-racist critique because we need to understand how it is that the world takes shape by restricting the forms in which we gather.... We need this critique now, if we are to learn how not to reproduce what we inherit.

And this form of critique is important to conduct because citation politics have been acknowledged but not thoroughly explored in the context of OOO. Indeed, our argument here and throughout this chapter is that ontologies themselves are inseparable from the scholarly practices that build and maintain them. As Annemarie Mol (2002) argues, "*ontology* is not given in the order of things, but that, instead, *ontologies* are brought into being, sustained, or allowed to wither away in common, day-to-day, sociomaterial practices" (p. 6). As feminist thinkers have long recognized, scholarly work is not a referential medium that represents some external reality; rather, such work is what participates in and enacts those worlds. Ontologies are thus bound up with and enabled by the most mundane of practices; and in the case of OOO, an object-oriented understanding of the world is predominantly enacted through scholarly work taking place in both print and online forums.

We want to do more than point out this problem in the hopes of better understanding how scholarly practices enact ontologies. Citations, we might say, not only help us build worlds but build worlds populated by people who have been and will continue to be active participants in them. And citations do this by serving various rhetorical purposes. We might cite in order to acknowledge the importance of a particular piece of scholarship and thus also meet the demands of unspoken disciplinary conventions that require us to demonstrate such knowledge to readers; to situate our work in the context of scholarly conversations that have previously engaged ideas on or related to our current inquiry; or, perhaps, to articulate an exigency motivating present concerns and the value these concerns may have for others. As Latour points out in *Science in Action* (1987), citations help scholars rhetorically form allies that aid in mobilizing credibility and strengthening the knowledge claims put forth in scholarly research.

But much is at stake in both our way of using citations and the way we understand our own citation practices. Haraway (1997), for one, takes issue with Latour's military vision of scholarly production in technoscience, arguing that approaching citations as a machination of war is an overtly masculinist and limited way of understanding what it means to reference another's work. As she describes,

> war is the great creator and destroyer of worlds, the womb for the masculine birth of time. The action in science-in-the-making is all trials and feats of strength, amassing of allies, forging of worlds in the strength and numbers of forced allies. All action is agonistic; the creative abstraction is both breathtaking and numbingly conventional. Trials of strength decide whether a representation holds or not. Period. (p. 34)

The strength of knowledge claims, in Latour's understanding, depends on how well they abide by spoken or unspoken conventions that help them form alliances with established scholarly work. Pointing out the dangers of such conventions, Haraway (1997) describes how women were excluded from the Royal Society of London's scientific reports: "Women might watch a demonstration; they could not witness it" (p. 31). That is, "women's names were never listed among those attesting the veracity of experimental reports, whether they were present or not," thus leading to the development of a "masculine scientific modesty" that would continue to be the hallmark of early historical accounts of scientific work (p. 32).

What Haraway's account makes clear is that the very mundane work of citation has powerful ontological effects in the world. Moreover, there is an important element of historical construction to consider here. Indeed, Charles Bazerman (1993) understands citations as "sites at which communal memory is sorted out and produced" (p. 20). Such communal memory, in the context of Haraway's modest witness, is shaped by the "conventions of modest truth-telling," through which the Royal Society's experiments allowed women to watch the demonstration but prohibited them from witnessing it: "The definitive demonstrations ... had to take place in proper civil public space, even if that meant holding a serious demonstration late at night to exclude women" (1997, p. 31). What makes such conventions so problematic—and so difficult to address—is how they often circulate and persist as unspoken or unacknowledged ways of doing things.

In critiquing mundane citation practices in OOO scholarship, we aim both to acknowledge OOO scholars' lack of engagement with feminist thought and reconceptualize scholarly work as having ontological consequences that warrant deeper investigation. That is, we do not simply want to critique OOO for failing to properly or adequately cite work by feminist scholars. Rather, we want to push the critique further by calling attention to the ontological impact that not citing feminist scholarship on materiality has in the world. In practice, the consequences of OOO scholars' ignoring feminist work include the infantalization of academic women, such as Haraway (see Harman, 2010, "Question Period"); the equating of bodies and objects with no contextualization or mention of possible effects; and the erasure and subsequent "inventing" of ideas from feminist theory. These consequences may not show up for philosophers, as they are often trying to define the world as it is, but they do show up for rhetoricians, who attempt to identify the effects of different ways of being in the world. Whether or not OOO scholars want to be aware of the consequences of their work, those consequences still exist. Further, the question may not be whether OOO should acknowledge these sources but whether OOO can be a viable methodology or line of inquiry while ignoring the vast majority of feminist work that engages the same philosophical terrain.

"Girls [Women] Welcome!!!" on Matters of Accountability

The lack of female voices in the OOO movement is not a secret. Practitioners of OOO often seem genuinely perplexed as to why more women aren't interested or don't (seem to) participate in conversations about objects. In live-blogging a talk given by Haraway (2010), Harman writes, "Not enough girls in speculative realism which makes her [Haraway] mad, but she's still curious and seduced by it. [Note: Girls welcome!!!]" (2010, "Question Period"). Assuming the dismissive way he summarizes Haraway's point is due not to a lack of respect but rather the informality of the medium, "Girls welcome" is still somewhat troubling. The word choice itself suggests the call's lack of seriousness, though perhaps what's more worrisome is the way public calls for inclusion (such as actual citation and engagement) can be used to disguise exclusionary practices. Sara Ahmed (2013) writes:

> We begin with a friendly openness. It's an open call, they say. Come along, they say. Take our places, they even say. Note here how the gesture of inclusion, which is also a promise of inclusion, can be offered in a way that negates a point about exclusion. To suggest incorporation as potential (come along as you *can* come along) blocks an acknowledgement that the open call was restricted as a call.

Harman's public call masks that the actual scholarly practices he and others in OOO engage in exclude female voices; it also fails to acknowledge that the call to talk about objects is not his to make. Indeed, such talk has been the focus of feminist scholarship for years. OOO has already drawn a line in the sand on the beach of objectivity, putting women on one side and OOO on the other, despite what Harman's public invitation may suggest.

In *Women, Science, and Technology: A Reader in Feminist Science Studies* (Wyer, Barbercheck, Cookmeyer, Öztürk, & Wayne, 2014), the theme of scientific objectivity is pervasive throughout. Contributors continually critique methods of knowing that posit absolutes: "Institutions of higher education—their development, organization, practices, and underlying assumptions—have inherited the Enlightenment legacy of (white) male-as-objective, with troubling consequences" (p. xx). That's not to say OOO is yet another instantiation of this legacy, but rather that the desocializing, decontextualizing, depoliticizing, and deethicalizing of theory works to reinstitute the idea that things, including ideas, can be objective. The way OOO "cleans up" its objects by removing their ties, contexts, and histories is troubling and has ontological consequences.

We opened this chapter with a discussion of trends associated with object-oriented philosophy, OOO, and object-oriented rhetoric, and we worried about the potentially dangerous scholarly practices that exist below the surface in much OOO scholarship. We talked about the erasures of feminist thought, technoscience, and new materialist work from OOO's theorizing of objects, and we argued that these erasures are part of what makes OOO a potentially unethical way to practice scholarship. In this section, we briefly review a few of the scholars who have been most markedly erased from the masculine resurgence of interest in objects, scholars who have long been doing work "discovered" by practitioners of OOO, but doing it in a cautious way that considers the ethical implications of scholarly work.

FRSS, the topic of this book, is a kind of perpetually emergent, rather than canonical, collection of writers and writings. In other words, FRSS is a category that can be used to describe some fundamental questions and ways of questioning in primarily feminist scholarship that have been happening since the 1970s. Concepts such as the posthuman, embodiment, ecologies, materiality, objects, nature and animals, contextualization, and scholarly practices have historically been at the center of what feminist scholars think and write about. Scholars interested in these categories have emerged from many disciplines, making FRSS truly interdisciplinary in its thought and methodologies. The idea that fundamentally defines OOO (as opposed to SR, as described by Harman)—that nonhuman objects have agency and engage in interaction without human interference or even knowledge—has been well theorized, discussed, and eventually problematized by the very scholarship OOO has, for the most part, ignored.

OOO's claim that objects are always withdrawing and have meaning outside of their interactions with humans is both interesting and legitimate, if not a bit simplistic. Listing and juxtaposing objects has a certain shock value that shakes scholars from our language- and human-centered perspectives. If OOO scholars had looked to, or chosen to acknowledge, the widely acknowledged predecessors of that ontological line of questioning, however, they would have found a rich body of scholarship that expands, explores, and eventually explodes this idea. In this scholarship, done primarily by feminist researchers in and of science, the goal is not to raise the status of objects to the level of the human, nor is it to lower the human to the status of an object. Indeed, such a goal is often quickly rejected by feminist science studies (FSS) scholars (as described in our following discussion of both Karen Barad and N. Katherine Hayles) because they refuse to make such sharp distinctions between objects and humans. To raise the ontological status of an object to that of a human asserts that there is some sort of demarcation, or solid boundary, between them. Scholars of FSS are altogether not convinced this is the case.

For example, in her seminal work, *Meeting the Universe Halfway: Quantum Physics and the Entanglement of Matter and Meaning* (2007), Barad writes that

> phenomena—whether lizards, electrons, or humans—exist only as a result of, and as part of, the world's ongoing intra-activity, its dynamic and contingent differentiation into specific relationalities. "We humans" don't make it so, not by dint of our own will, and not on our own. But through our advances, we participate in bringing forth the world in its specificity, including ourselves. (p. 353)

Throughout her work, Barad argues that objects, humans, space, and time are ontologically inseparable. We all "intra-act," and these intra-actions bring about our very ontological being.

The difference between intra-action and OOO concepts discussed earlier may seem subtle, but the nuance here actually makes all of the difference. Rather than erasing bodies, ignoring politics and ethics, and sidestepping the absolute critical importance of context, intra-action makes these things part of something's ontology: "The spacetime manifold does not sit still while bodies are made and remade. The relationship between space, time, and matter is much more intimate" (Barad, 2007, p. 376). This means that every atom, every moment, and every context is part of an object's ontology. Any definition of object that does not account for ethics, then, is suspect: "Objectivity is a matter of accountability for what materializes, for what comes to be. It matters which cuts are enacted: different cuts enact different materialized becomings" (p. 361). To explain on the quantum level why time and ethics intra-act with objects outside of human intervention is beyond our technical expertise and the scope of this chapter. We urge those interested in the promises of OOO to read Barad's work more closely. What is significant here, though, is that we cannot be so myopic in our treatment of objects or humans, as all atoms are intra-acting together to bring particular things forward and obscure others. Erasing the researcher's values, beliefs, and feelings is not only impossible but unethical. Talking about HIV, toads, and an eraser as if we as researchers get to instill neutrality into an object that has its own politics and ethics is a dangerous fantasy.

Barad provides a philosophical as well as a physical justification for why objects/subjects/time/ethics cannot be separated:

> A delicate tissue of ethicality runs through the marrow of being. There is no getting away from ethics—mattering is an integral part of the ontology of the world in its dynamic presencing. Not even a moment exists on its own. "This" and "that," "here" and "now," don't preexist what happens but come alive with each meeting. The world and its possibilities for becoming are remade with each moment. If we hold on to the belief that the world is made of individual entities, it is hard to see how even our best, most well-intentioned calculations for right action can avoid tearing holes in the delicate tissue structure of entanglements that the lifeblood of the world runs through. (p. 396)

The world is entangled. Objects matter. Objects matter outside of their relations with humans. However, advocating for a view of HIV as being no more important than the discarded rind of an orange ignores nearly everything that makes HIV HIV and a rind a rind, except perhaps that they both exist, which to our knowledge was never really in question.

Another rich area that has been sidestepped in OOO is the discussion of the posthuman in FSS, particularly the posthuman body. N. Katherine Hayles (1999) productively problematizes the way our bodies function and exist in a posthuman world:

> The posthuman view configures human being so that it can be seamlessly articulated within intelligent machines.... [T]here are no essential differences or absolute demarcations between bodily existence and computer simulation, cybernetic mechanism and biological organism, robot teleology and human goals. (p. 3)

This does not put the body on a higher plane of existence than an object. On the contrary, it refuses to posit an absolute demarcation between them. As seen throughout much of Hayles's work as well as in the work of many FSS scholars, this supposition is anything but arbitrary. It shows that information, ideas, and perceptions are embodied within artifacts. This, along with physicality, is what gives artifacts and objects an undeniable agency.

This argument does not privilege the body, as OOO may caution, but it recognizes that the "human being is first of all embodied being, and the complexities of this embodiment mean that human awareness unfolds in ways very different from those of intelligence embodied in cybernetic machines" (Hayles, 1999, p. 284). This recognition of the complexity, interconnectedness, and history of the body is unsettlingly absent in theories of OOO. Hayles further argues that "meaning is not guaranteed" (p. 285). So, while there is considerable shock value in listing "things" as being on equal ontological footing, these "things" have contexts, histories, and agencies in ways that create different meanings, meanings to humans and meanings to other objects. FSS is not content to stop at the argument that things matter, as that has never been enough for scholars like Hayles. Rather, FSS scholars push beyond those simplistic questions of existence and ontology toward questions of boundedness and meaning.

While Barad's work is somewhat more recent (2007), the work of Hayles and Haraway is not. That said, while both have spoken on the subject of either OOO or SR, we find it deeply troubling that they are so rarely cited in OOO scholarship despite their decades of theorizing materialism, objects, and relations between humans and the environment. In her seminal article, "Situated Knowledges: The Science Question in Feminism and the Privilege of Partial Perspective" (1988), Haraway cautions:

> One cannot "be" either a cell or a molecule—or a woman, colonized person, laborer, and so on—if one intends to see and see from these positions critically. "Being" is much more problematic and contingent. Also, one cannot relocate in any possible vantage point without being accountable for that movement. Vision is *always* a question of the power to see—and perhaps of the violence implicit in our visualizing practices. (p. 585)

To declare that an ice cube has the same ontological weight as a rape ignores not only the utter privilege and domination wielded by the scholar who declares it but also the violence that domination does to others (both humans and objects).

In *The Companion Species Manifesto* (2003), Haraway further underlines her position that no ontology can be severed from another: "Reality is an active verb.... Beings do not preexist their relatings" (p. 6). She talks about beings "co-constituting" each other (p. 32). This co-constitution is not merely a philosophical exercise; rather, as we see in both Haraway and Barad, objects and humans are continually being changed by one another on a quantum level. This interconnectedness is acknowledged over and over in the work of FSS scholars, but it is conspicuously absent from OOO. Or perhaps said more fairly, the ontological weight of these connections seems to have been ignored. Haraway cites 315 women out of 558 citations in *Simians, Cyborgs, and Women: The Reinvention of Nature* (1991). These are connections that matter, and these are connections that have ontological and ethical consequences. As Barad (2007) argues, ethics is impossible to separate from matter, so much so that toward the end of her book she refers to matter as "ethics-matter."

It is this notion of "ethics-matter" that provides the starting place for our methodology, which we discuss in the next section.[2]

A Methodology of Reverberations

In *Feminist Rhetorical Practices: New Horizons for Rhetoric, Composition, and Literacy Studies* (2012), Royster and Kirsch call for a new methodological perspective for feminist rhetorics. Their intent is to develop a way for feminist rhetoricians to engage in "rhetorical assaying," whereby scholars learn to "focus on inquiry strategies and on sorting through the impacts and consequences" of their scholarly work (p. 42). One of the terms of engagement Royster and Kirsch offer is "strategic contemplation," which they define as a deliberate form of reflection: "Taking the time, space, and resources to think about, through, and around our work as an important meditative dimension of scholarly productivity" (p. 21). We understand Royster and Kirsch's notion of strategic contemplation as a methodological attitude that values slow, reflective, and cautious thinking. As they describe it, strategic contemplation

> means recognizing what was made possible for us as feminist rhetorical scholars through other women's work, how their efforts have enabled us to stand where we are today, and how their visions make it possible for us to imagine a future worth working for. (pp. 22–23)

Taking our cue from Royster and Kirsch, we propose a feminist rhetorical methodology that is likewise committed to consciously and intentionally reflecting on our scholarly work and the ontology it enacts in the world. That is, we see a need for a way to attend carefully and cautiously to the ontological impact and consequences of our scholarly practices, including how we cite the work of others. Following Patricia Sullivan and James Porter (1997), we understand methodology to mean the driving theoretical frameworks of knowledge and experience that guide specific methods or practices. These frameworks, or terms of engagement, help foreground the political and ethical importance of taking the time to think through what ontologies we enact in our most mundane of practices. From this methodological perspective, ontology is not only the focus of scholarship but is very much enacted in the work we do. We advocate, in other words, for a shift from imagining ontology as strictly a topic of inquiry to exploring *how we enact ontologies through practice.*

To do so, we develop a methodology grounded in the notion of reverberations. Reverberations, as a concept, appeals to us because it conveys a sense of lasting and continuing effects that seem to emanate from a designated origin.

Here, we propose that scholarship also reverberates—it makes an impact and that impact ripples outward. Rhetorical reverberations, in other words, help us gauge how our decisions in scholarship will affect past scholarly work as well as how it will initiate future conversations. If we understand scholarship as an ontological enactment, reflecting on the reverberations of our scholarship can guide future scholarly work toward more ethical ends. In many ways, our use of reverberations here works toward understanding the ontological impact of the rhetorics we build and the histories we maintain.

Our methodology is indebted to numerous feminist scholars who have come before us. What we propose has been largely influenced by Annemarie Mol's *The Body Multiple: Ontology in Medical Practice* (2002), which argues for an ontological politics whereby ontology is not predicated on existing states of the world but is instead enacted in day-to-day life. Similarly, we advocate for remaining critical of and attentive to the important relationship between ontologies and methodologies. Indeed, John Law (2004), echoing a long tradition of social scientists, science and technology studies scholars, and feminist theorists, argues that "methods, their rules, and even more methods' practices not only describe but also help to *produce* the reality that they understand" (p. 5). Such a reminder, Law notes, is not meant to reintroduce notions of the modest witness who continuously strives to remain outside of the situation he or she is studying. It is, in Law's words, not a recommendation of "political quietism" but rather proceeds from the notion that

> since social (and natural) science investigations interfere with the world, in one way or another they always make a difference, politically and otherwise. Things change as a result. The issue, then, is not to seek disengagement but rather with how to engage. (p. 7)

We hold that when ontologies are concerned, it is best to engage scholarly work by continually acknowledging the way it reverberates, both into the past (by constructing a history of indebtedness to previous scholarship) and into the future (by reflecting on the social and political impact our work may have on others).

Reverberations into the Past: On Citation Practices

The reverberations of scholarly work not only shape the future but also construct a comprehendible narrative of the past through our practices of citation. Scholars from rhetoric, composition, professional and technical communication, and education have long recognized the rhetorical significance of citations in written work (Bazerman, 1987, 1993; Connors, 1998, 1999; Howard,

1995; (In)Citers, 1998; Rose, 1993). Every citation, for Robert Connors (1998), "implies a universe of meanings . . . [and] declares allegiances and counterallegiances" (pp. 6–7). Moreover, many have specifically foregrounded the importance of citations in cultivating interdependence and dialogue among diverse perspectives (Jung, 2014; Powell, 2004; Yood, 2013). Such interdependence, or indebtedness, has likewise been at the core of a wide range of feminist studies scholarship (Ahmed, 2013; Haraway, 1994; Healy, 2015; Hemmings, 2011; hooks, 1984; McDermott, 1994; Smith, 1981). We situate our discussion of reverberations into the past in this previous work in order to describe the salient problems of OOO's construction of a philosophical history that excludes feminist thought. As an integral form of scholarly work, citations are concrete practices that foreground the boundary-drawing work of scholarship, and they make apparent how reverberations into the past circulate and rhetorically shape how future scholarly work is done.

Our methodology of reverberation, we suggest, emphasizes the political and ethical importance of engaging the vast wealth of previous work that has come before us. It consciously recognizes our knowledge and work as always indebted to ways others have made sense of the world. But indebtedness is less a passive state than it is an attitude that motivates critical examination of how we are constructing knowledge and where bridges may be built to deepen our work. Haraway (1994) offers the game "cat's cradle" as a trope that places our collective indebtedness at the center of an analytical practice:

> Cat's cradle is about patterns and knots. . . . One person can build up a large repertoire of string figures on a single pair of hands; but the cat's cradle figures can be passed back and forth on the hands of several players, who add new moves in the building of complex patterns. (pp. 69–70)

The indebtedness of cat's cradle is not a rule or requirement of the game but what makes it all the more interesting. Moreover, like our mundane scholarly practices, what is at stake is not immediately apparent: "if we do not learn how to play cat's cradle well, we can just make a tangled mess," but, Haraway continues, "if we attend to scholarly, as well as technoscientific, cat's cradle with as much loving attention as has been lavished on high-status war games, we might learn something about how worlds get made and unmade, and for whom" (p. 70). Similarly, understanding the reverberations of our indebtedness is one way for us to key into how mundane scholarly practices, such as citations, help build and maintain worlds that others might find uninhabitable.

Our methodology of reverberations, as a habit of mind that foregrounds the importance of citation practices, helps us attend to points of convergence between one's own work and various other disciplinary perspectives. What this most often means is that we must become involved in the messy work of taking apart our own ideas, tracing underlying connections and influences, and ethically acknowledging the voices that emerge. Considering the reverberations of how we cite those to whom we are indebted is a slow and careful process. It can never guarantee that all direct or indirect exclusions have been identified, but as a habit of mind it can keep these concerns at the foreground of one's inquiry. We can attend to our indebtedness by asking a series of critical questions: How is this inquiry contextualized? In what ways are the questions that drive this work predicated on previous knowledge? Who else has asked similar questions? Who else has similar concerns? What history is being constructed? What narratives are being articulated, and who has been forgotten? What effects will this inquiry have on the material world? Critical questions along these lines are most effective when they are built into our practices of scholarly work, rather than tacked on as an afterthought.

Coincidentally, Levi Bryant (2014) has recently addressed the issue of citation politics in a way that resonates with what we are proposing as rhetorical reverberations. In a blog post titled "The Politics of Distinctions/The Politics of Citation" (2014), Bryant argues that citations always involve distinctions that are statements of value indicating notions "of what is worth thinking, of who is worth attending to, of who is worth hearing." Most often, he acknowledges,

> we encounter the criticism that you're not asking the questions and doing the research in the area the respondent would like to do, as if everyone is able to say everything and do everything.
>
> Of course, while that style of criticism is often uncharitable—"why aren't you talking about what I want to talk about?"—there is a real set of political issues here. Why are these questions, problems, issues, and themes being put on the table and not others?

Bryant is careful to attend to the systemic problem underlying this "uncharitable" style of criticism. In order to remain charitable in our own critique of OOO and citation politics, we quote Bryant's concluding point in full:

> Often the failure to cite is just plain ignorance of this work. However, that explanation will only take us so far because we have a duty to acquaint ourselves with [work] being done in our area. Oversight itself is, as it were, a product of distinctions. Yet again, today it is impossible for anyone to master all that has been written, even if ones [sic] own subject area. The archive has truly become Borgesian. The challenge then is that of how we

can balance charity or the recognition of human finitude with our duty to cite and recognize the convergent work of others. However, the issue is not as simple as this, for when we look at patterns of citation—especially in a discipline like philosophy—we often see that it does not seem to be simple oversight that leads the work of other thinkers not to be cited, but that there appears to be a more fundamental, far more political, set of distinctions at work functioning to select who is cited and who is not cited.

As we have discussed thus far, the issues involved in failing to engage and cite feminist thought are more often the effects of scholarly practices and less so the result of individual ignorance. Bryant describes these concerns in terms of a scholarly "duty to engage with that research that is relevant to and convergent with our own." But as is clear from OOO scholarship, such engagement is not always grounded in a conscious sense of individual duty; rather, it is often woven into the mundane and often invisible practices that undergird scholarly work.

Much of Bryant's post addresses the concerns we have outlined in this chapter. Moreover, his insistence that citational exclusions are not always intentional and malicious resonates with our sense of reverberations, which are not always within our control. We are, in many ways, always indebted, and thus always responsible for reflecting on the histories we are building and those we have forgotten along the way. Indeed, Bryant's blog post advocates for the kind of methodology we propose here:

> We shouldn't begin with the premise that the person has *malicious* intent in distinguishing as they do. While they do indeed use the distinction, that distinction is invisible to them. This is why critical work revealing distinctions that underly [sic] a particular form of indication are valuable.

Reflecting on the reverberations of citation practices and the way such practices signal our indebtedness to those who came before us helps foreground the need for reflective approaches to citations as a necessary and ethical part of our scholarly practices.

Reverberations into the Future: On Reflective Practices

While the social and political impacts of our citation practices reverberate into the past by constructing intellectual histories, our scholarly work likewise affects future conversations and scholarship. Our methodology of reverberations not only reflects on the past but also considers the possible social and political impact our scholarly work may have in the future. A number of strategies for doing so are available, such as Latour's (1997) notion of "shifting out and shifting in" or Clifford Geertz's (1973) "tacking in and tacking out." Similarly, Katie King (2012) offers a productive tactic that she calls "scoping and scaling" that resonates with our understanding of reverberations. As she describes it, scoping and scaling allow us to

> see the whole territory [as] we pan out and up for a satellite view, or we come in closer and closer to see the very particular street patterns, maybe even to detail the backyard of a specific house, the parking lot of a particular building. We move the orientation point around with our mouse, cursor, finger, or whatever, to shift *scope and scale*. (p. 4)

Adapting this as a heuristic guideline for scholarly work, we might imagine scoping and scaling as a scaffolded approach to stepping back from a project and surveying the larger theory or history one is constructing. Moreover, scoping and scaling ask us to see beyond the borders of our scholarly work and speculate on what rhetorical effect it might produce in the world.

Cultivating this kind of sensitivity to our work's rhetorical reverberations can begin by asking ourselves a series of critical questions: How will the contextualization of my work affect readers from various disciplines? How will it affect those who have pursued similar lines of inquiry? What assumptions have been left unaddressed in the questions driving this work? Who have I not considered that has also addressed this topic? How might they respond to the way I have described previous scholarship and the approach I take in doing this work? In what ways will the theory or history being developed in this work directly or indirectly harm others?

As some of these questions make clear, pursuing an understanding of rhetorical reverberations means speculating about the future; it also often means going back and tracing an idea and gaining a sense of how the development of that idea has obscured some element that may produce an unintended effect. Giorgio Agamben's (2009) work on method articulates this kind of attitude as an "archaeological vigilance" through which our inquiry must "retrace its own trajectory back to the point where something remains obscure and unthematized" (p. 8). Rhetorical reverberations as a habit of mind means continuously looking back and looking forward to get a better sense of what one's scholarly work will do in the world.

Our understanding of rhetorical reverberations is not antithetical to OOO. In fact, much of what we mean by it resonates with Bogost's (2012) notion of alien phenomenology: "Just as the astronomer understands stars through the radiant energy that surrounds them, so the philosopher understands objects by tracing their impacts on the surrounding ether" (p. 33). Where alien phenomenology differs from our sense of reverberations is in its commitment to speculating about object-to-object relations. "As philosophers," Bogost writes,

> our job is to amplify the black noise of objects to make the resonant frequencies of the stuffs inside them hum in credibly satisfying ways. Our job is to write the speculative fictions of their processes, of their unit operations.... Our job is to go where *everyone* has gone before, but where few have bothered to linger. (p. 34)

As scholars of FRSS, we see our methodology as a way of tracing the social and political impact of our scholarly work on others in the world.

OOO's collective use of the word *object*, rather than *thing* or *materiality*, offers a salient example of where a consideration of scholarly work's rhetorical reverberations may be valuable. *Thing*, as Bogost describes, is a term with a "troubled philosophical history" (so much so that Bogost, in fact, prefers *unit* to both *object* and *thing*) (p. 24). Traditionally, for Heidegger, an everyday object becomes a thing when one experiences a breakdown in its use. Heidegger's favorite example, of course, is the hammer, which only becomes foregrounded in our minds when it breaks. While Heidegger's account of the hammer as thing evinces his preference for things above objects, Latour (2005) taps into the etymology of *thing*, which highlights the term's original reference to "a certain type of archaic assembly" where many would gather to discuss matters of concern (p. 22). It is this sense of things—as a "dingpolitik" that gathers humans and nonhumans into a collective—that has rendered *thing* a term that, according to Bogost, "carries considerable baggage" in connecting objects solely in relation to human subjects (2012, p. 24). Things, in other words, work rhetorically to bring into readers' minds a certain conception of the material world—the deliberate use of the term reverberates and is perceived to have tangible effects on how others understand it. The same, we suggest, is true of objects. This is most clear in Harman's (2013) suggestion that "the term 'objects' is not opposed to 'subjects,' so it is not such a bad fate to be an object" (p. 39). As scholars of FRSS will quickly attest: it is, in fact, a bad, even terrible, fate to be an object. While the basic principles of OOO aim to raise the ontological status of objects to people, not to lower people to the ontological status of objects, the deliberate use of the term *object* reverberates and means different things to different people. This is particularly important when considering marginalized groups that have continually struggled with oppressive regimes that consciously or unconsciously undercut their claim to existence.

One possible objection here might be that ontology sees no gender, race, or sexual identity. Indeed, Bogost (2012) says as much in the concluding section of *Alien Phenomenology* when he writes: "Being is unconcerned with issues of gender, performance, and its associated human politics; indeed, tiny ontology invites all beings to partake of the same ontological status" (p. 99). However, similar appeals have been made time and time again that present scientific inquiry as a neutral arbiter of fact that has no concern for politics. Furthermore, *object* as a term exhibits a rhetorical force that cannot be ignored. In fact, Bogost describes being confronted with this reality at the opening of an OOO symposium held at Georgia Tech in the spring of 2010. As he describes it, the symposium's website was designed to present an image of a different object each time users visited the page. The page may, on one visit, exhibit an array of rusty nuts and bolts and, on another visit, show a neatly stacked tower of comic books. The algorithm that drew these images from Flickr, however, presented an unexpected problem for Bogost and other symposium organizers when "a (female) colleague had showed the site to her (female) dean—at a women's college, no less. The image that apparently popped up was a woman in a bunny suit" (p. 98). While Bogost and Bryant never saw the exact image, they understood that the website aligned OOO with a sense of objectification. Bogost goes on to admit that

> given the charged nature of the subject—a sexist "toy" on a website about an ontology conference organized by and featuring 89 percent white men—it would have been tempting to shut down the feature entirely or to eviscerate its uncertainty and replace it with a dozen carefully suggested stock images, specimens guaranteed not to ruffle feathers. (pp. 98–99)

Bogost was forced to weigh the benefits of the website's deliberate intention to align its design with OOO's basic ontological principles against the risks of users misinterpreting the philosophical movement: "As anyone who has used the Internet knows all too well, the web is chock-full of just that sort of objectifying images exemplified by the woman in the bunny suit" (p. 99). Finally, Bogost notes, the symposium organizers edited the code for the search query:

> Options.Tags="(object OR thing OR stuff) AND NOT (sexy OR woman OR girl)." (p. 99)

The problem, as we see it, concerns more than the wealth of objectifying images circulating on the Internet on a daily basis. The problem is also—and perhaps more so—the fact that searches for *object* result in images of woman at all. In such cases, to be an object is not simply a bad fate; it is a fate tragically entangled with the continual reminder that women have historically been seen as nothing more than objects. Placing "objects" at the center of one's ontology, then, demands that one acknowledge this reality with great caution and care.

Conclusion

We hold that reflecting on the rhetorical reverberations emanating from our scholarship can help scholars gain a better understanding of how ontologies are built and maintained by the most mundane of practices. The methodology we propose is not, in any sense, prescriptive but is instead intended to be a heuristic that provides a way of integrating critical reflections into our scholarly work. Alongside this methodological perspective, one of the driving goals of this chapter has been to acknowledge the long history of work by feminist thinkers who have continued to account for the complex ways the material world is made and unmade. Our purpose here in many ways echoes Haraway's intent in *Modest_Witness@Second_Millennium. FemaleMan©_Meets_OncoMouse™* (1997):

> I want feminists to be enrolled more tightly in the meaning-making processes of technoscientific world-building. I also want feminists—activists, cultural producers, scientists, engineers, and scholars (all overlapping categories)—to be recognized for the articulations and enrollments we have been making all along. (p. 127)

We too hope to include the richness of feminist thought on the material world in the histories told and retold by new materialist scholars, particularly proponents of OOO.

In a much more general sense, we have argued throughout this chapter that the practices used in the production of scholarly work enact particular ontologies. We have contended that a feminist methodology of engagement can help us understand that ontologies are first and foremost built and enacted by scholars and have veritable effects in the world. Our methodology calls for new materialist scholars to critically consider the mundane practices through which these ontologies are established and maintained. Doing so can draw our attention to the social and political complexities that are ever-present in scholarly endeavors. More specifically, we have come to see that productive and equitable engagement in discussions of the material world calls for a sustainable way of thinking and talking about ontology that keeps the social and political implications of our scholarly work at the forefront of our minds. What is more, we have also come to understand that ontology is best approached not as a disciplinary terrain on which to fight but as an ongoing ethical practice. That is to say, scholarly work concerning the nature of what is real is never a purely intellectual production; rather, such work is best understood as a rhetorical account of the world that acknowledges and affirms past and present voices as well as sustains reflection on what our work does to those who have long struggled with cultural narratives that have told them time and time again they are objects and nothing more. Such an accounting, enacted in the methodology we propose, is in no way incompatible with the array of new materialist philosophies that have been the topic of recent academic conferences, journal articles, and books. Rather, our methodology simply, but forcefully, asserts that in increasingly complex discussions of the material world, we must sustain our efforts to open spaces for more complexity, caution, and care.

Notes

1. Harman cites Lynn Margulis and Sarah Ruel-Bergeron; Bryant cites Karen Barad, Jane Bennett, Claire Parnet, Donna Haraway, N. Katherine Hayles, Susan Oyama, Isabelle Stengers, and Virginia Woolf; and Bogost cites Jane Bennett, Suzanne Cotter, N. Katherine Hayles, Gina Macris, Susan Schulten, Christy Lange, Fiona Candlin, and Christine Kamprath.

2. While we have not named (nor could we ever name) the many FSS scholars who discuss these issues, we hope to provide compelling evidence that crucial contributions by feminist scholars have been erased from and overlooked in current discussions of ontologies and objects. Indeed, our discussion of three scholars (Barad, Hayles, and Haraway) offers only a glimpse into the complexity of FSS that we cannot do justice to in this chapter. That said, we believe that the absence of these three prominent and prolific scholars from OOO suggests either a kind of willful exclusion, or else an utter lack of awareness of anything labeled "feminist."

References

Agamben, G. (2009). *The signature of all things: On method*. (L. D'Isanto & K. Attell, Trans.). Cambridge, MA: Zone Books.

Ahmed, S. (2013, September 11). Making feminist points [Web log post]. *Feministkilljoys*. Accessed 20 June 2014. Retrieved from http://feministkilljoys.com/2013/09/11/making-feminist-points/

Barad, K. (2007). *Meeting the universe halfway: Quantum physics and the entanglement of matter and meaning*. Durham, NC: Duke University Press.

Barnett, S. (2010). Toward an object-oriented rhetoric [Review of the books *Tool-being: Heidegger and the metaphysics of objects* and *Guerrilla metaphysics:*

Phenomenology and the carpentry of things, by Graham Harman]. *Enculturation, 7.* Accessed 8 January 2013. Retrieved from http://enculturation.net/toward-an-object-oriented-rhetoric

Bazerman, C. (1987). Codifying the social scientific style: The APA *Publication Manual* as a behaviorist rhetoric. In J. S. Nelson, A. Megill, & D. N. McCloskey (Eds.), *The rhetoric of the human sciences: Language and argument in scholarship and public affairs* (pp. 125–144). Madison: University of Wisconsin Press.

Bazerman, C. (1993). Intertextual self-fashioning: Gould and Lewontin's representations of the literature. In J. Selzer (Ed.), *Understanding scientific prose* (pp. 20–41). Madison: University of Wisconsin Press.

Bogost, I. (2011, December 26). Object-oriented answers [Web log post]. *Bogost.* Accessed 20 June 2015. Retrieved from http://bogost.com/writing/blog/object-oriented_answers/

Bogost, I. (2012). *Alien phenomenology, or what it's like to be a thing.* Minneapolis: University of Minnesota Press.

Braidotti, R., & Vermeulen, T. (2014, August 12). Borrowed energy. *Frieze, 165.* Accessed 1 August 2015. Retrieved from https://frieze.com/article/borrowed-energy

Brown, B. (2001). Thing theory. *Critical Inquiry, 28*(1), 1–22.

Bryant, L. R. (2010, June 29). Vitale on SR and politics [Web log post]. *Larval Subjects.* Accessed 28 April 2015. Retrieved from https://larvalsubjects.wordpress.com/2010/06/29/vitale-on-sr-and-politics/

Bryant, L. R. (2011). *The democracy of objects.* Ann Arbor, MI: Open Humanities Press.

Bryant, L. R. (2011, December 22). Some responses to Jussi [Web log post]. *Larval Subjects.* Accessed 28 April 2015. Retrieved from https://larvalsubjects.wordpress.com/2011/12/22/some-responses-to-jussi/

Bryant, L. R. (2014, October 19). The politics of distinctions/the politics of citation [Web log post]. *Larval Subjects.* Accessed 28 April 2015. Retrieved from https://larvalsubjects.wordpress.com/2014/10/19/the-politics-of-distinctionsthe-politics-of-citation/

Butler, J. (1990). *Gender trouble: Feminism and the subversion of identity.* New York, NY: Routledge.

Butler, J. (1993). *Bodies that matter: On the discursive limits of "sex."* New York, NY: Routledge.

Connors, R. J. (1998). The rhetoric of citation systems, part 1: The development of annotation structures from the Renaissance to 1900. *Rhetoric Review, 17*(1), 6–48.

Connors, R. J. (1999). The rhetoric of citation systems, part 2: Competing epistemic values in citation. *Rhetoric Review, 17*(2), 219–245.

Coole, D., & Frost, S. (Eds.). (2010). *New materialisms: Ontology, agency, and politics.* Durham, NC: Duke University Press.

Cooper, M. M. (1986). The ecology of writing. *College English, 48*(4), 364–375.

DeLanda, M. (2002). *Intensive science and virtual philosophy.* New York, NY: Continuum.

Derrida, J. (1988). *Limited Inc.* Evanston, IL: Northwestern University Press.

DeVoss, D. N., Cushman, E., & Grabill, J. T. (2005). Infrastructure and composing: The *when* of new-media writing. *College Composition and Communication, 57*(1), 14–44.

Dolphijn, R., & van der Tuin, I. (2012). *New materialism: Interviews and cartographies.* Ann Arbor, MI: Open Humanities Press.

Edbauer, J. (2005). Unframing models of distribution: From rhetorical situation to rhetorical ecologies. *Rhetoric Society Quarterly, 35*(4), 5–24.

Geertz, C. (1973). *The interpretation of cultures.* New York, NY: Basic Books.

Haas, C. (1996). *Writing technology: Studies on the materiality of literacy.* New York, NY: Routledge.

Halberstam, J. (2012, June 15). MIA to IA [E-mail list message]. *Empyre.* Accessed 22 April 2015. Retrieved from http://lists.artdesign.unsw.edu.au/pipermail/empyre/2012-June/005276.html

Haraway, D. (1988). Situated knowledges: The science question in feminism and the privilege of partial perspective. *Feminist Studies, 14*(3), 557–599.

Haraway, D. (2003). *The companion species manifesto: Dogs, people, and significant otherness.* Chicago, IL: Prickly Paradigm Press.

Haraway, D. (2010). *Response to Isabelle Stengers.* Paper presented at the Fourth International Conference of the Whitehead Research Project, Claremont, CA.

Haraway, D. J. (1991). *Simians, cyborgs, and women: The reinvention of nature.* New York, NY: Routledge.

Haraway, D. J. (1994). A game of cat's cradle: Science studies, feminist theory, and cultural studies. *Configurations, 2*(1), 59–71.

Haraway, D. J. (1997). *Modest_witness@second_millenium. FemaleMan©_meets_OncoMouse™: Feminism and technoscience.* New York, NY: Routledge.

Harman, G. (2002). *Tool-being: Heidegger and the metaphysics of objects.* Peru, IL: Open Court.

Harman, G. (2005). *Guerrilla metaphysics: Phenomenology and the carpentry of things.* Peru, IL: Open Court.

Harman, G. (2009). *Prince of networks: Bruno Latour and metaphysics.* Melbourne, Australia: re.press.

Harman, G. (2010). *Towards speculative realism: Essays and lectures.* Washington, DC: Zero Books.

Harman, G. (2010, December 3). Question period: Stengers and Haraway on speculative realism [Web log post]. *Object-Oriented Philosophy.* Accessed 9 May 2015. Retrieved from http://doctorzamalek2.wordpress.com/ 2010/12/03 /question-period/

Harman, G. (2011, December 23). Jussi Parikka on OOO [Web log post]. *Object-Oriented Philosophy.* Accessed 28 April 2015. Retrieved from https:// doctorzamalek2.wordpress.com/2011/12/23/jussi-parikka-on-ooo/

Harman, G. (2013). *Bells and whistles: More speculative realism.* Washington, DC: Zero Books.

Hayles, N. K. (1999). *How we became posthuman: Virtual bodies in cybernetics, literature, and informatics.* Chicago, IL: University of Chicago Press.

Healy, K. (2015, February 25). Gender and citation in four general-interest philosophy journals, 1993–2013 [Web log post]. *Kieran Healy.* Accessed 1 June 2015. Retrieved from https://kieranhealy.org/blog/archives/2015/02/25 /gender-and-citation-in-four-general-interest-philosophy-journals-1993 -2013/

Heidegger, M. (2008). *Being and time* (Reprint ed.). (J. Macquarrie & E. Robinson, Trans.). New York, NY: Harper Perennial Modern Thought.

Hemmings, C. (2011). *Why stories matter: The political grammar of feminist theory.* Durham, NC: Duke University Press.

hooks, b. (1984). *Feminist theory: From margins to center.* New York, NY: South End Press.

Horner, B. (2000). *Terms of work for composition: A materialist critique.* Albany: State University of New York Press.

Howard, R. M. (1995). Plagiarisms, authorships, and the academic death penalty. *College English, 57*(7), 788–806.

(In)Citers. (1998). The citation functions: Literary production and reception. *Kairos, 3*(1). Accessed 8 June 2015. Retrieved from http://kairos.technorhetoric .net/3.1/binder2.html?coverweb/ipc/authorship.htm

Jung, J. (2014). Systems rhetoric: A dynamic coupling of explanation and description. *Enculturation, 17*(1). Accessed 9 October 2016. Retrieved from http://www.enculturation.net/systems-rhetoric

King, K. (2012). *Networked reenactments: Stories transdisciplinary knowledges tell.* Carbondale: Southern Illinois University Press.

Latour, B. (1987). *Science in action: How to follow scientists and engineers through society.* Cambridge, MA: Harvard University Press.

Latour, B. (1993). *The Pasteurization of France.* (A. Sheridan & J. Law, Trans.). Cambridge, MA: Harvard University Press.

Latour, B. (1997). Where are the missing masses? The sociology of a few mundane artifacts. In W. E. Bijker & J. Law (Eds.), *Shaping technology / building society: Studies in sociotechnical change* (pp. 225–258). Cambridge: Massachusetts Institute of Technology Press.

Latour, B. (2005). From realpolitik to dingpolitik, or how to make things public. In B. Latour & P. Weibel (Eds.), *Making things public: Atmospheres of democracy* (pp. 1–32). Cambridge: Massachusetts Institute of Technology Press.

Law, J. (2004). *After method: Mess in social science research.* New York, NY: Routledge.

McDermott, P. (1994). *Politics and scholarship: Feminist academic journals and the production of knowledge.* Urbana: University of Illinois Press.

Mol, A. (2002). *The body multiple: Ontology in medical practice.* Durham, NC: Duke University Press.

Morton, T. (2012). *The ecological thought.* Cambridge, MA: Harvard University Press.

Morton, T. (2013). *Hyperobjects: Philosophy and ecology after the end of the world.* Minneapolis: University of Minnesota Press.

Parikka, J. (2011, December 21). OOQ—object-oriented-questions [Web log post]. *Machinology.* Accessed 1 December 2015. Retrieved from https://jussiparikka.net/2011/12/21/ooq-object-oriented-questions/

Powell, M. D. (2004). Down by the river, or how Susan La Flesche Picotte can teach us about alliance as a practice of survivance. *College English, 67*(1), 38–60.

Reid, A. (2013). What is object-oriented rhetoric? [Audio file]. *Itineration.* Accessed 16 September 2014. Retrieved from http://tundra.csd.sc.edu/itineration/node/11

Rice, J. E. (2008). Rhetoric's mechanics: Retooling the equipment of writing production. *College Composition and Communication, 60*(2), 366–387.

Rickert, T. (2013). *Ambient rhetoric: The attunements of rhetorical being.* Pittsburgh, PA: University of Pittsburgh Press.

Rose, S. K. (1993). Citation rituals in academic cultures. *Issues in Writing, 6*(1), 24–37.

Royster, J. J., & Kirsch, G. E. (2012). *Feminist rhetorical practices: New horizons for rhetoric, composition, and literacy studies.* Carbondale: Southern Illinois University Press.

Smith, L. C. (1981). Citation analysis. *Library Trends, 30*(1), 83–106.

Star, S. L. (1995). Introduction. In S. L. Star (Ed.), *Ecologies of knowledge: Work and politics in science and technology* (pp. 1–38). Albany: State University of New York Press.

Sullivan, P., & Porter, J. (1997). *Opening spaces: Writing technologies and critical research practices.* Westport, CT: Ablex.

Syverson, M. A. (1999). *The wealth of reality: An ecology of composition.* Carbondale: Southern Illinois University Press.

Wyer, M., Barbercheck, M., Cookmeyer, D., Öztürk, H. Ö., & Wayne, M. (Eds.). (2014). *Women, science, and technology: A reader in feminist science studies* (3rd ed.). New York, NY: Routledge.

Wysocki, A. F. (2004). Opening new media to writing: Openings and justifications. In A. F. Wysocki, J. Johnson-Eilola, C. L. Selfe, & G. Sirc, *Writing new media: Theory and applications for expanding the teaching of composition* (pp. 1–42). Logan: Utah State University Press.

Yood, J. (2013). A history of pedagogy in complexity: Reality checks for writing studies. *Enculturation, 16*(1). Accessed 25 May 2015. Retrieved from http://enculturation.net/history-of-pedagogy

2.

Flat Ontologies and Everyday Feminisms

Revisiting Personhood and Fetal Ultrasound Imaging

Jen Talbot

While I was writing this, the U.S. Fourth Circuit Court of Appeals declared a law—the Display of Real-Time View Requirement—unconstitutional; it required that physicians performing abortions in North Carolina must not only perform a mandatory ultrasound but also place the ultrasound image in the patient's line of vision and provide a detailed verbal description of the image (*Stuart v. Camnitz*, 2014). This provision, and others like it that have proliferated in recent years, provides the occasion to revisit the role of fetal ultrasound technologies in constituting the fetal subject, placing it in relation to pregnant and maternal subjects and considering how we navigate relationships among ontology, ethics, and the law. Definitions of personhood, which underwrite proposed legislation like the Display of Real-Time View Requirement (hereafter, the Display Requirement), are philosophically and legally contested, in part because *personhood* and *agency* are slippery concepts.

"Personhood," for example, is a legal status that is rooted in the characteristics, rights, and obligations of being a specific, individual human who acts in the world. In other words, personhood as a concept is based on the idea of the rational and self-interested humanistic subject (Langford, 2015). In the United States, personhood is synonymous with citizenship, in that the fundamental rights associated with personhood status and granted by the Constitution apply to all "born or naturalized" citizens. Bodily autonomy has historically served as both a condition for and a right granted by personhood status. In other words, legal personhood and the rights thereof follow for persons whose ontological status as such is unquestioned. However, unquestioned recognition of full personhood through the practical recognition of legal rights has historically been dependent on a social understanding of embodied subjectivity. The fundamental rights of legal personhood have not always been (and in many cases are not) a given for all people, based on bodily attributes and perceived bodily attributes such as gender and gender presentation, race, and ability. In fact, antiabortion rhetoric of the Reagan era explicitly tied *Roe v. Wade* (1973) to the *Dred Scott* decision of 1857, in which the court ruled that African Americans were not citizens and therefore not granted the fundamental right to freedom guaranteed by the Constitution (Condit, 1990, p. 50). *Dred Scott* was overturned by the Thirteenth and Fourteenth Amendments, which abolished slavery and expanded civil rights, respectively.

The history of the legal category of personhood has been one of increasing inclusivity and thus generally representative of the core progressive and feminist values of equality, fairness, and diversity. The attribution of personhood to nonhuman entities can be viewed as an extension of this inclusivity. This is what makes the case of fetal personhood such an interesting one, as it creates a tension in which the adoption of generally held feminist values—the expansion of equality and the recognition of diverse enactions of agency—undermines the personhood of women, in that women's bodies become subject to public and state surveillance and control.[1] As a concept, then, personhood demonstrates the inseparability of the ontological and the rhetorical: fetal personhood and its conflict with the practical enactment of women's personhood demonstrate the inseparability of the bodily and the social. Fetal ultrasound technologies are of perennial interest for feminist rhetorics because they offer a stark illustration of how crucial the social and embodied are to the constitution of the subject and the enactment of agency; further, they demonstrate the key role of the phenomenological in building ethical rhetorics and politics.

Using the Display Requirement as an example of the general increase in fetocentric legislation in recent years, this chapter traces how fetal agency is differentially enacted from the perspective of two different posthumanist theories—Bruno Latour's Actor-Network Theory (ANT) and Karen Barad's agential realism. Specifically, I consider the increasing ontological weight ascribed to fetal subjects, a shift that was facilitated by ultrasound technologies and intensified by the legal discourses surrounding their use. I argue that agential realism offers a feminist posthumanist framework that takes seriously lived experience and the ways embodied forces (affective, biochemical, and neurological) influence emotions, relationships, and behaviors. By conceptualizing embodiment as the phenomenological ground and medium through which the social emerges, agential realism offers a complex theory of distributed accountability that takes into consideration the lived experiences of others. As such, it has greater potential to reconcile asymmetries among human persons ethically and compassionately while still working toward extending personhood beyond the human. Ultimately, I argue that posthumanist feminisms, by virtue of their deep engagement with the body and material influences on it, provide a means to integrate nonhuman agencies into rhetorical theory and practice, while maintaining concern for the ways social asymmetries act on the bodies of people through lived experience.

To make this argument, I begin by identifying its exigency: namely, the need to address tensions that arise when posthumanist conceptualizations of the social become entangled with feminist politics.

Posthumanisms and Critiques of the Social

Many iterations of feminism, like most social justice orientations, tend toward the humanistic: a belief in rational subjects applying their will to persuade others to work toward positive change and a more egalitarian world. Though the humanist subject requires an epistemic rather than an ontological version of rhetoric, as well as a view of agency that is often a double-edged sword, it also provides affordances for accountability and resistance that many posthumanist rhetorics do not. In general, posthumanist thought, and its particular strains such as ANT, object-oriented ontology, and feminist materialism, share the central tenet that human beings are shaped by nonhuman agents to the same extent that entities in the world are shaped, managed, and driven by humans. Posthumanist theories that posit a flat ontology acknowledge a symmetrical distribution of agency among humans and nonhumans: agency and affectability are recognized as distributed more evenly throughout networked ecologies of humans, animals, technologies, spaces, and things that are motivated by material, discursive, and affective forces.[2] An entity can effect change—be an agent of action—without being a sentient or self-aware subject, a shift that complicates humanistic categories of personhood. Agency is thus defined by effect rather than intention. It is neither a characteristic that a particular entity has nor a form of currency that can be spent to intentionally influence people or situations; rather, it exists only in interactions—the result of a confluence of forces that come together to create particular effects in ways that are sometimes the result of intentionality and sometimes not. The intersection of these forces is of interest to a wide variety of scholarly disciplines, given that a posthuman perspective can and often is viewed as an ethical remedy to the Cartesian perspective that casts humankind as the center, purpose, and primary actor of life.

The attention to materiality that makes and enables a posthumanist ontology grew as a response to scholars' overreliance on human-centric, social constructivist theories of culture and discourse to describe phenomena. Latour's (2005) ANT, for example—which aims to observe and map associations among agential forces (both human and nonhuman)—is a critique of sociology's traditional methodologies, especially with regard to their study of science and technology. In the framework that Latour works against, "the social" is an amorphous entity and domain that acts as a backdrop for material interactions. His reenvisioning of the social brings two ideas to the forefront: first, that inquiries into the social are no longer bound to the scale of human phenomenological experience; and second, that the social is neither a kind of "glue" that holds nonsocial things together nor is it a variety of things that are held together by some other means. For Latour, "the social" is a term that too often provides a "free ride" to observers seeking connections among actants—agential entities that create effects by connecting with other actants.[3] The social, as an explanation, covers over structures that might, with the introduction of different actants, configure themselves differently. In order to avoid this, ANT attempts to "render the social world as *flat* as possible in order to ensure that the establishment of any new link is clearly visible" (p. 16). Connections between actants at the micro- and macrolevels (individual and network, respectively) and among unstable frames of reference take the place of human-centered attempts to stabilize any particular social frame.

For feminist rhetoricians, the flattened ontologies characteristic of posthumanism are problematic when they render illegible the historical and political forces that shape our everyday experiences. Additionally, the collapse of the category of human and the placing of it within a constellation of nonhuman agents can erase the agential asymmetries that exist among human bodies. In other words, while the inclusive ethics of much posthumanist theory appeals to feminist principles, its tendency to ignore existing asymmetries within the categories of human and person does not.

Although a humanist version of feminism, with its focus on human agency and autonomy, affords more obvious avenues for addressing these asymmetries, the connection between it and feminist rhetoric is tenuous. We can specifically observe this in appeals to individual freedoms and bodily autonomy that surround law and policy relevant to reproductive rights. *Roe v. Wade* was a judgment on the right to privacy; the Display Requirement was struck down because its compulsory speech constituted a First Amendment violation. In such cases, the rhetoric of personal autonomy serves feminist ends: it argues that if personal and therefore bodily autonomy is what we value in legal and political realms, then that value should be applicable to women as well as men. However, personal autonomy is not necessarily a tenet of feminist rhetoric, which tends to place greater emphasis on interdependence, distributed agency, and situatedness. For this reason, a feminist rhetoric of personal autonomy in the context of establishing and defending reproductive freedoms is best understood as tactical: it is not an appeal to individualism as an abstract philosophical value; rather, it is a response to the promise of bodily autonomy that has historically not been upheld for women.

As assumptions about individual freedom and discrete human agency are challenged, however, the efficacy of this tactical rhetoric wanes and the rhetorical foundations of reproductive freedom in the United States are destabilized: practices of posthumanist life gather intensity and become increasingly codified into infrastructure and law, and the assemblage of a woman's body, a fetus, ultrasound technology, law, and media effect a new ontological status for the fetus. Indeed, when the fetal body is included in an expanded notion of personhood, the pregnant woman and fetus become ontological equals whose interests are, in some cases, in conflict. Monica Casper (1998) straightforwardly articulates the tension that emerges when feminism, posthumanism, and reproductive rights meet:

> I want historically "nonhuman" people and animals to have agency (and I must admit I worry less about machines in this regard), but I do not necessarily want fetuses to have agency. . . . My refusal to grant agency to fetuses, while simultaneously recognizing it in pregnant women and in my cats, is about taking sides. My politics . . . are about figuring out to whom and what in the world I am accountable. (pp. 852–853)

In Casper's case, accountability is specifically a sociopolitical notion: it is the inverse of agency—a state in which an entity is influenced or acted on—that is combined with a clear and subjective intentionality such that agency becomes visible (or not) in accordance with an a priori set of ends in mind. I argue that posthumanist orientations that are feminist in both theoretical origin and methodology productively theorize the kind of accountability Casper describes. By offering rich and multidimensional understandings of affect, embodiment, and the biology of emotion and persuasion, posthumanist feminisms offer possibilities for imagining rhetorical interventions that neither humanist feminisms nor ANT can provide. Furthermore, I argue that posthumanist frameworks that minimize the phenomenological, such as Latour's ANT, are problematic for feminist rhetorics, since human bodies' experiences of the social (for example, practices that sustain white privilege) drive human action. For feminist rhetoricians, then, any approach that ignores how social asymmetries are reinforced and replicated on the level of embodied human experience is not a viable option.

To make this argument, I draw on work by materialist feminists such as Karen Barad (1998, 2007), Rosi Braidotti (2006), Donna Haraway (1991, 1997), and Susan Hekman (2008), who share Latour's skepticism of the utility of social constructivism. However, they differ from Latour in that the lived experience of the body remains a central concern and that the material body is contiguous with the material world. The microscale interactions between body and world are the mechanisms by which social forces manifest in the body over time; in the same instance, however, these are the very material interactions that Latour argues are covered over by the social. This attention to the material does not mean that the human scale is considered a stable boundary that defines an agential being. On the contrary, in Barad's (2007) agential realism the basic ontological unit is a *phenomenon*, not a body. As she explains,

> the primary ontological units are not "things" but phenomena—dynamic topological reconfigurings/entanglements/relationalities/(re)articulations of the world. And the primary semantic units are not "words" but material-discursive practices through which (ontic and semantic) boundaries are constituted.... Agency is not an attribute but the ongoing reconfigurings of the world. (p. 141)

In other words, what we experience as the agential body is the negotiated confluence of microscopic interactions between organic and inorganic, human and nonhuman, material and discursive, chemical and electrical, and psychic elements. All these elements work together to project the phenomenologically integrated human body that acts, interacts, makes decisions, and collaborates with other entities and agents to create social effects.

For Barad, Haraway, and other feminist materialists, the social emerges from, but cannot be reduced to, the material body, and connections between agents are felt and inscribed into the physiology, biochemistry, and habituated responses of the body. For example, the long-term emotional and behavioral effects of trauma or the epigenetic effects of poverty are material inscriptions of social phenomena. By contrast, Latour's limited attention to the body allows him to dismiss asymmetries among social groups that are largely created and maintained through different legal and cultural orientations toward the bodies of those identified with different social groups, particularly in terms of race, gender, and class (Sturman, 2006). Latour (2005) argues that the "flagrant asymmetry of resources does not mean that they are generated by social asymmetries," and in fact the presence of asymmetries is evidence that factors other than the social are in play (p. 64). While it is rare that a particular asymmetry would be *solely* due to social forces, Latour does not address the ways in which technoscientific practices in large part constitute many social inequalities (Haraway, 1997; Sturman, 2006). Specifically, ANT seeks to identify and describe the interactions of all manner of animal, technological, and discursive networks such that "each participant is treated as a full-blown" agent of action (Latour, 2005, p. 128), yet it categorically ignores significant ways of filling gaps between agents through affective and embodied means, as articulated by feminist materialists.

For example, although Latour's stated argument is that the social and material are empirically inseparable (particularly in macro-interactions in which bodies and objects act as networks, with consequences that exceed the intention of the individual), his focus is on the agencies of things and the ways in which things constitute assemblies and forces that have been shorthanded as "social." However, equal attention is not given to the converse: the ways social categories are written on material bodies through the biochemistry of affective and emotional interactions as well as economic and environmental factors such as nutrition, safety, the biochemistry of stress and trauma, and so on. From this perspective, humans are moved not only by virtue of being part of agential networks larger than themselves but also by biological and chemical agencies much smaller than (and constitutive of) the embodied self. These micro-actants function as mediators of human behavior within networks, though they are often not readily identified as causes. In feminist visions of posthumanist interactions, the body is the mediator of social forces in a very real (if not always visible) way. In short, feminist materialist theories foreground the inseparability of the social and material through the lived experience of micro-interactions.

In subsequent sections, I explore this inseparability by examining the interaction of biology, technology, discourse, law, and ethics as configured by the Display Requirement. The Display Requirement provides a useful case with which to draw out the nuances of different iterations of posthumanist philosophies, in that it brings together the proliferative theoretical possibilities of technoscience and ontology and puts them in relationship with the concrete enactments necessitated by law. In other words, like ontological politics as a concept, the Display Requirement "[has] to do with the way in which 'the real' is implicated in the 'political' and *vice versa*" (Mol, 1999, p. 74). Through this case study, I aim to illustrate the practical implications of ANT and agential realism, respectively, through their different conceptions of the body's role in constituting the subject. While the two approaches share many similarities and the differences between them are subtle in theory, the effects of their differences are considerably more dramatic.

Legal Provisions and Fetocentrism

In the Display Requirement, which was originally enacted as part of 2011's Women's Right to Know Act (the rest of which remains in practice), the ultrasound is a mandated requirement for informed consent to terminate a pregnancy. Specifically, the law attempted to mandate that a woman seeking an abortion be subject to both the ultrasound image and a detailed verbal description of the image (the woman could not refuse to undergo the ultrasound, although she could avert her eyes and cover her ears during the description). As we will see in this chapter, the narration of the ultrasound procedure serves an important role in drawing the fetus into social networks. The rhetorical act of narration is second only to the ultrasound technology itself in terms of assembling the network surrounding the fetus.

A number of feminist theorists, scientists, and rhetoricians have taken up questions raised by the increasing ubiquity and sophistication of fetal ultrasound technologies and how they affect the personhood status of fetuses and women. Judith Butler (1993), Karen Barad (1998), Monica Casper (1994, 1998), and Charlotte Kroløkke (2011), for example, approach the question from the techno-ontological angle, while Celeste Condit (1990), Lisa Mitchell (2001), Carol Sanger (2008), Catherine Langford (2015), and Katie Gibson (2008) foreground the rhetorical, legal, and social aspects of what is, ultimately, a complex apparatus that limits reproductive freedoms and influences the trajectories of women's lives. As Gibson argues, "fetal pain research and 3-D ultrasound technologies have provided pro-life advocates new 'scientific' warrants to advance the personhood of the fetus" (2008, p. 329).

The scientific argument for fetal personhood is further strengthened when it is codified into law. It is worth noting that the Display Requirement was declared unconstitutional because it violates free speech, but the court also noted that the law risks portraying pro-life ideology as an undisputed medical fact. Specifically, the court ruled that the Display Requirement violates the First Amendment rights of the technician: the state cannot compel a technician to advocate a particular position in regard to the ontological status of the fetus because the technician has individual autonomy over whether or not to speak in a given situation. In its ruling, the court opined that the patient would also be in danger of misinterpreting medical speech as state-compelled ideological speech:

This Display of Real-Time View Requirement explicitly promotes a pro-life message by demanding the provision of facts that all fall on one side of the abortion debate—and does so shortly before[4] the time of decision when the intended recipient is most vulnerable. (2014)

The conflation of medical knowledge and state-compelled ideology is reinforced by the ways in which supporters of the law (and others like it) have redefined the notion of informed consent, which does not simply provide a patient with necessary information about a medical procedure; it also forces her to visually confront a particular moral interpretation of her actions. Encoding fetal ontology into law thus changes the notion of informed consent from a medical necessity to the acceptance of an entire constellation of moral consequences, as well as to a particular social narrative of maternal identity (Sanger, 2008). In other words, the presence of the fetus not only indicates the medical diagnosis of pregnancy but also forces the pregnant woman into the social category of "mother," with the moral accountability that goes along with it.

Legislation like the Display Requirement, which collaborates with efforts to garner legal personhood for the fetus, is thus constituted through the mandated and strategic use of ultrasound technology. For example, as of December 2014, twelve states[5] required that women be provided with written or verbal information on accessing ultrasound services; in nine states,[6] if the ultrasound was performed as preparation for an abortion procedure, the provider had to offer the patient the opportunity to view the image; in an additional ten states,[7] prior to the abortion, the provider had to offer the patient the opportunity to view the image; in three states[8] (now, after the decision in North Carolina, it is two), the provider had to display and describe the image to the patient. More recently, bills that define a fetus as an autonomous legal person with rights, in some cases from the moment of fertilization, have been put forth in Colorado (twice, in 2008 and 2010), Missouri, Nevada, Oklahoma, and Virginia. Similarly, fetal heartbeat bills, which prohibit abortion after detection of a fetal heartbeat, have been attempted in Alabama, Arkansas, Kansas, Kentucky, Mississippi, North Dakota, Ohio, Texas, and Wyoming. All these provisions depend on technologies that make fetal presence perceptually available to parties other than the pregnant woman, at a much earlier stage in development than has historically been possible.

Prior to the twentieth century, quickening, or the first fetal movement detectable by the pregnant woman, was the legal standard for considering a fetus as a separate entity (Condit, 1990; Gibson, 2008). In a first pregnancy, this typically occurs between eighteen and twenty weeks and can happen a few weeks earlier in subsequent pregnancies. Prior to the development of ultrasound technologies, then, the ontological status of the fetus was confirmable only via the pregnant woman's sensory perceptions, her interpretation of movement as fetal movement, and her willingness to communicate those perceptions to others. Though the fetus is an agent of material changes to the maternal body prior to this point, many of those changes are imperceptible or illegible as the effects of pregnancy. Fetal movement is typically discernable to others between twenty-four and thirty weeks, at which point the pregnancy is most likely visually noticeable as well.

Two-dimensional ultrasound imaging, which became increasingly common for low-risk pregnancies during the 1980s and was part of standard prenatal care by the 1990s, is typically performed in order to determine a fetus's sex[9] around the same time as quickening. In this case, sensory confirmation from an outside source is possible at around the same time that it is possible for the mother. Three-dimensional ultrasound imaging can visually identify sex as early as eleven weeks, although, interestingly, it is often not recommended until the third trimester, so the fetus looks "more like a baby," or "less like an alien" or "more human." While 3-D ultrasounds are considered displayable as early baby pictures for family and friends or even broadly sharable via social media, it is not the image's detailed representation that raises questions about the fetus as agent (questions that are rendered even more difficult by technologies that make a twenty-four-week-old fetus a viable baby if born prematurely); rather, the fetus's status as agent depends on the timing of the ultrasound, specifically, how that timing positions the fetus, the state, and the pregnant woman in relation to one another.

Approximately 90 percent[10] of abortions occur in the first trimester (with one-third of the total number occurring within the first six weeks, at which time the embryo is measurable in millimeters), and very little that is recognizable can be seen with transabdominal (external) ultrasound, regardless of whether it is 2-D or 3-D. A transvaginal ultrasound, which requires that the pregnant woman be penetrated vaginally, and which was the primary component of the controversial 2012 mandate in Virginia, can reveal a visible heartbeat as early as six weeks.[11] In legislation like the Display Requirement, sensory verification of the fetus's ontological presence, previously available only to the pregnant woman (and, perhaps, others at her invitation), is now mandated without her consent. Additionally, the ultrasound technician is in the position to mediate (through the rhetorical act of narration) the agential interaction between pregnant woman and fetus. In the case of the pregnant woman, perceptual access to the fetus is the result of the fetus's physiological and biochemical enactment of agency; for the technician, perceptual access is visual, and the fetus is visually represented as a discrete entity. If we ignore the ways in which agencies have effects on the body, this combination of visual and rhetorical representation displaces the pregnant woman as primary mediator of fetal agency and replaces her with, in the case of the Display Requirement, an agent of the state.

The pregnant woman's role as mediator of fetal personhood was, at the time of *Roe v. Wade*, so uncontroversial that the fact that the court agreed to hear the case automatically excluded the possibility of fetus as person. As Celeste Condit (1990) points out,

> [accepting the case] allowed women to bring pregnancy (perhaps our single most important issue as women) into the judicial process in a definitive way. Clearly, fetuses could not bring such cases, and adults had been routinely denied the right to bring such cases for fetuses who were not eventually born. Thus, by merely hearing the case, the Court granted women a crucial form of social power and standing that fetuses could not have. (p. 100)

Invoking Pierre Bourdieu, Catherine Langford (2015) similarly argues that the occupation of social space is a necessary condition for personhood: "Within constitutional law the fetal body, occupying the woman's physical space, does not secure a separate social space—that of <person>hood, with all the rights and liberties secured thereby—until it appropriates a physical space separate from her" (p. 136). In both of these cases, the agency of personhood (in Condit's example, specifically the rhetorical agency granted by personhood) is predicated on physical presence in a social space. In other words, because a fetus is not a discrete physical entity with an autonomous social presence, it cannot be granted legal standing.

Yet the distributed agency enabled by the Display Requirement and constitutive of posthumanist ontologies opens up the conceptual possibility of fetal personhood. When agency—as a constitutive property of personhood—is detached from single autonomous bodies to include assemblages that are both larger and smaller in scale, significant grounds for valuing the personhood of women over that of fetuses are eroded. In previous discussion, my analyses of the role of ultrasound technologies on fetal personhood were primarily based on the perceived need (from an antiabortion perspective) to "speak for" the fetal person, whose ontological reality is constructed on visual confirmation by an outside party. In the following section, I will analyze more fully the construction of fetal personhood in the context of the Display Requirement—and, importantly, its implications for feminist politics—using two different posthumanist frameworks—ANT and agential realism, respectively.

ANT and the Display Requirement

As discussed earlier, Latour's ANT is a methodology for tracing new connections and assemblages as they form. Flat ontologies identify new associations occurring within a particular phenomenon (Latour, 2005, p. 11). Because it emphasizes associations enabled by new technologies, ANT is particularly well suited to tracing the ways that the increasing clarity and ubiquity of fetal ultrasound technology create denser networks to mediate fetal agency. The Display Requirement in particular draws discursive and legal constituents into a network that had been previously constituted by primarily biological, medical, and technological actants. Flattening out the social world, ANT renders both the pregnant woman and the fetus actants in a shared network in which the political, ethical, and social asymmetries that exist prior to the ontological presence of the fetus are irrelevant (p. 11). In a flat ontology, then, the pregnant woman and the fetus are ontologically and agentially equivalent in a situation that does not take into consideration the ways in which women's bodies are and have been mediated, constrained, and inscribed differently from men's bodies within an organized but asymmetrical system of distributed agency.[12]

Indeed, Latour has been criticized for conceptualizing the body as just another actant in networks and ignoring how the body is acted on differently as a result of social/rhetorical categories (Haraway, 1997; Sturman, 2006). This is in part a result of the ongoing tension between ontology and politics, but it is also a demonstration of rhetoric's need, in its engagement with posthumanism, to retain a phenomenological element. In "How to Talk about the Body?" (2004), Latour defines the body as *"an interface that becomes more and more describable as it learns to be affected by more and more elements"* (p. 206). For Latour, being more affected is always better than being less affected: it is the means by which one becomes an "articulate subject," one who has the capacity to create ever-increasing distinctions through progressively having a body that is acted on (p. 210). The process of learning to be affected he describes through the example of training to be "a nose":

> Starting with a dumb nose unable to differentiate much more than "sweet" and "fetid" odours, one ends up rather quickly becoming a "nose" (*un nez*), that is, someone able to discriminate more and more subtle differences and

> able to tell them apart from one another, even when they are masked by or mixed with others. It is not by accident that the person is called "a nose" as if, through practice, she had *acquired* an organ that defined her ability to detect chemical and other differences. Through the training session, she learned to have a nose that allowed her to inhabit a (richly differentiated odiferous) world. Thus body parts are progressively acquired at the same time as "world counter-parts" are being registered in a new way. Acquiring a body is thus a progressive enterprise that produces at once a sensory medium *and* a sensitive world. (pp. 206–207)

By conceptualizing the body as emergent, as capable of being acquired, Latour avoids a model in which there is a body (subject), a world (object), and an intermediary (language) that connects them. An intermediary, in sociology, is an entity that is part of a networked interaction but does not change the functioning of the network; it is an inert object through which force, will, or information passes (Latour, 2005). For Latour, the body-object-language model is problematic in part because it helps sustain divisions between object(tivity) and subject(ivity) that in turn authorize the hegemony of Science (2004, p. 222). He is also concerned with the ways in which the model renders elements like language and technologies "intermediaries," which fails to recognize their agential force while also maintaining the problematic divisions he seeks to dismantle. Rather than regard seemingly inert objects as intermediaries, Latour posits that they are *mediators*—actants in a network—with *network* defined as "a string of actions where each participant is treated as a full-blown mediator. To put it very simply: A good ANT account is a narrative or a description or a proposition where all the actors *do something* and don't just sit there" (2005, p. 128).

If we apply this logic to pregnancy, the fetus does not become an actant until it is perceived as doing something, and there are any number of ways this might happen. Annemarie Mol (1999), whom Latour cites in "How to Talk about the Body?" (2004), provides an analogy relevant to understanding the multiple ways a fetus becomes an actant. She describes three methods of identifying anemia, or, rather, three ways in which anemia is articulated. The first is clinical performance, in which a patient presents a set of visible symptoms—in this case, the color of the inner eyelid. The second is laboratory performance, in which anemia presents hemoglobin levels below a statistical standard. The third is pathophysiological, in which hemoglobin levels have dropped below the levels necessary to circulate oxygen through the body of a particular person (p. 78). Performances of anemia via each of these channels may or may not coexist. For example, a patient might present hemoglobin levels that fall below the statistical threshold but are still sufficient to carry oxygen throughout the body; or, a patient might present no visible symptoms but not be getting sufficient oxygen. This multiplicity of diagnoses results in multiple bodies, coexisting within one another at a given moment.

Similarly, as a woman's body is affected by the presence of the fetus, a number of diagnostic methods may be used at different times and with varying degrees of accuracy. Prior to the nineteenth century, some biochemically feasible methods did exist (for example, mixing wine with urine, in which case proteins in the urine would react with the alcohol); nevertheless, a sexually active woman's best option for diagnosing pregnancy was awareness of her own cycle and potential early symptoms (National Institutes of Health, 2003). In these cases, the biological and chemical effects of pregnancy render the fetus an agent via the subjective embodied experience of the woman. With the development of tests for human chorionic gonadotropin, a hormone produced by the placenta at implantation, a woman in the 1930s and 1940s was increasingly encouraged to have her self-diagnosis of pregnancy confirmed by her doctor, which distributed mediating agency beyond the woman to include medical personnel as well. This introduction of additional agents into the network weakened the ontological hierarchy of woman and fetus created by the physical dependency of the fetus and introduced the proposition of the fetus as person into social space. With the development of home pregnancy tests in the mid-1970s, however, this diffusion of mediating agency into a social space was reversed: pregnancy was confirmed not by a doctor but by a test taken at home (National Institutes for Health, 2003). The medical category of pregnant woman and the social category of mother began to blend together.

Each of the aforementioned methods of diagnosis yields a different articulation of subject and world. In the context of Latour's project of challenging the certainty of scientific objectivity, these multiple articulations are desirable, since they extend controversies and maximize disputability; they also offer a patient the opportunity "to become affected" and thus a more complexly articulated subject (2004, p. 225). Despite its affordances for science studies, however, the concept of a disputable body introduces a number of political concerns involving bodily autonomy, which Mol (1999) and Strathern (1996) engage more directly than Latour. Most significantly, given my project, is the fact that multiple methods of diagnosis displace the patient as primary mediator and therefore locus of choice.[13] As Mol (1999) argues, different methods of diagnosis serve to "shift the *site* of the decision elsewhere" (p. 80). In the context of her analysis, the relevant decisions are treatment decisions, and the sites to which these decisions are displaced include measurement techniques and health-care budgets. In the case of the Display Requirement, the site of decision-making reconvenes in the medical-legal apparatus, which defines not only the woman as pregnant but also the fetus as a baby that exists within a set of social relationships. Simply put: regardless of the benefits of posthumanist theories that give rise to multiple bodies, *the law* considers—and acts on—a singular body. As a legal mandate, the use of ultrasound technology in the Display Requirement both increases disputability *and* shifts the location of choice away from the pregnant woman toward the state.

The key to this shift is the role of the technology itself as mediator, the agential force of which differs in relation to various ontological weights ascribed to the fetus. In a medical context, particularly in preparation for termination of a pregnancy, ultrasound images are used to determine things such as the size, location, and gestational age of the fetus, and to assess the level of risk to the woman undergoing the procedure. In these cases, the ultrasound imaging is used to see the fetus in the same way that ultrasound imaging might be used diagnostically or in preparation for any procedure that requires a visual confirmation of the location and state of subcutaneous soft tissue.[14]

The network in this case, then, involves a set of core agents: patient, technician, ultrasound technology, image, fetus. However, the mediating force of the technology is relatively weak: though it does bring the ultrasound technician and other medical personnel more tightly into the agential network, it is fairly close to being an intermediary. The system is moderately closed and the information disclosed is minimal and utilitarian: the technician views the image, and the pregnant woman does not (in fact, prior to the introduction of provisions like the Display Requirement, a woman's request to view the image would be refused). The technician shares very little, if any, of the information gathered from the image. Gestational age, for example, might be disclosed insofar as it would determine the procedure used for termination. The agents in the network can thus be traced as follows: the technician and technology act as intermediaries for the medical establishment/practitioner, gathering relevant information at the behest of the pregnant woman and making medical decisions without sharing the details of the ultrasound image. From the perspective of the pregnant woman, the fetus and ultrasound image are black-boxed into appearing as a single agent, the technician. The details of the procedure are based on fetal characteristics made visible by the ultrasound and interpreted by the technician, but the steps in this diagnostic process are not articulated either visually or verbally to the woman. Though the fetus is the agent of biological changes to the woman's body, the woman is, on a larger scale, the initiator of the technomedical actions that take place, and those actions are controlled in order to keep the pregnant woman in her role as primary mediator of the ontological presence of the fetus. In other words, by initiating the procedure, the woman has already made a decision in which the fetus is not a subject.[15] The roles of the other agents in the network are then aligned with this intention. In this case, then, the ontological benefit of multiplying bodies is subordinated to the ultrasound's diagnostic purpose as defined by the lived experience of the pregnant woman.

The mediating force of the ultrasound technology is greater in cases in which the ultrasound is part of the prenatal care of a pregnant woman who has taken on the role of mother and has thus initiated the ontological shift from fetus to baby. In such cases, the same set of agents—woman, technician, ultrasound technology and image, fetus—interact; however, the mediating force of the ultrasound is greater because it is part of the prenatal care for a pregnant woman who has taken on the role of mother. There are multiple occasions over the course of a full-term pregnancy to perform an ultrasound, which I outline in more detail later. In general, although the primary purpose of ultrasounds is diagnostic, they also serve several other purposes as well: to allow the mother to experience the fetus visually and aurally, to allow people other than the mother to experience the ontological reality of the fetus without the mediation of the mother, and to imbue the fetus with characteristics. The key point in all these occasions is that the pregnant woman's experiential response to her pregnancy drives the different enactments of the network, which purposefully operates to cast the fetus as a subject—a human person with a distinct identity who is capable of intentional action—rather than just an agent, an entity that influences its environment.

The earliest ultrasound generally does not provide a recognizably baby-like image, but the heartbeat is often visible and described in various reassuring ways—"strong," "healthy," and so on. Much like the shared experience of being able to feel the baby kick from outside, the early ultrasound allows for a shared experience of fetal ontological presence, one that can occur much earlier than without the benefit of ultrasound (or, in some cases, Doppler). As the pregnancy progresses, ultrasound can also tell the sex (based on genital appearance) of the baby, which of course carries a slew of attendant assumptions. In 1993, Judith Butler described ultrasound technology as a means of interpellating the fetal subject though gendering. Monica Casper (1994) also describes the making of the fetal subject, although she focuses on its construction as pediatric patient in cases of fetal surgery: the woman's abdomen and uterus are opened up in order to provide access to the fetus, which is operated on and returned to the uterus until birth. In these cases, while the woman is clearly also a patient, she is not the primary patient—in fact, the woman "may well be defined as part of the technology with which the FICU [fetal intensive care unit] is equipped" (p. 845).[16] The pregnant woman is therefore no longer the primary mediator of fetal presence, or even necessarily part of the mediating assemblage (in the same way as the ultrasound technology), but is instead a kind of life-support technology, similar to a ventilator or heart-lung bypass for an adult human. The pregnant body keeps the fetal body alive, but it is not necessary to the ontological status of the fetal patient. The interpellation of fetus as patient adds resonance to the overall interpellation of fetus as subject: it is another context in which fetal personhood and the personhood of the pregnant woman are disconnected from one another.

In addition to the interpellation of fetal subjects through gendering, or, much more infrequently, through the attribution of patient status, ultrasound technology also constructs the fetus as subject via a more casual assignment of traits based on the fetus's behavior. For example, the fetus's moving away from the pressure of the ultrasound wand might be described as "shyness"; being in a position that makes it difficult to visually determine the sex might indicate "stubbornness."[17] In these cases, the technician is doing more than guiding a wand and describing an image—he or she is contributing to a narrative of personality about the fetus, a narrative that would first be initiated by the mother in her role as mediator of the ontological status of the fetus. However, the personality narrative's construction is shared among the mother, the technician, and others who view the image in person or via social media after the fact. According to Charlotte Kroløkke (2011), this sharing and the subsequent construction of fetus as subject render the ultrasound experience a form of biotourism in which "sonographers take on the role of tourist guides and stage managers, and expectant parent(s) and guests are cast as tourists and coperformers" (p. 15). In other words, the ultrasound is an occasion for the fetus to be introduced as a social being, whose subjectivity is contained within but not mediated solely by the pregnant woman.

The fetus's ontological presence and perceived agency (and even, in cases of personality descriptors, subjectivity) are thus amplified by way of being shared among a greater number of people; the medical intervention makes this possible. In such a network, the fetus is an agent who is inscribed with personality and intentionality; the technician and technology are agents who participate in describing fetal characteristics to the parents/people present; and the human agents who are present to perceive this description will likely share this experience with others after the fact. Kroløkke (2011) calls this "the 'outing' of inner space," which is a means for constructing a "public citizen" (p. 15, 19). The fetus is not only a biological agent acting on the body of the pregnant woman, but it also becomes a visible subject, attributed with characteristics that categorize it into particular social groups as well as define it as an individual person, perhaps with a name, whose ontological and social presence is amplified and solidified as it is perceived by a greater number of mediators in the network. In these cases, the pregnant woman has intentionally drawn as many people as possible into a network that recognizes the personhood, the family-memberness, the social identity of her fetus. Indeed, the occasion of the fetal ultrasound is the point at which "the creation of the family album begins" (p. 15; Mitchell, 2001). The pregnant woman is still the primary mediator of fetal agency; however, through sharing that role with others, she also accepts her own inclusion into the social category of mother, a categorization that requires she consent to a state-defined moral accountability for her pregnancy.

In cases such as the Display Requirement and other legal maneuvers intended to establish legal personhood for the fetus, the medical, legal, and technological apparatuses that are deployed conflate the diagnostic medical purpose of ultrasound imaging and its social, emotional, and affective purposes. From the antiabortion perspective encoded within the Display Requirement, a pregnant woman seeking termination is consenting not only to a medical procedure but also to a particular narrative of fetal personhood and to the social category of mother (Sanger, 2008). As I will discuss more fully, the social category of mother implies full accountability for the well-being of the fetal person; the decision to terminate a pregnancy is interpreted as a failure of accountability not only to the fetus but also to the network of agents who make up the apparatus of personhood. The interests of the pregnant woman and the fetus are viewed as discrete and oppositional because the fetus is viewed as a person with interests and intentions that it is unable to communicate.

The move from fetal agent to fetal subject occurs at the point at which the pregnant woman, in interaction with medical, technological, and other human agents, initiates the process of providing sensory confirmation of the fetus for other people, who then have their own experiences of fetal presence, which are unmediated by the mother. The increasing visibility of nonhuman agents in everyday life, as well as public discourses about nonhuman persons, increases the ontological presence of the fetus not only as a particular case but also as a category of being. However, without the acknowledgment of fetal agency on the pregnant woman's body, emotions, and behavior, *any concept of the fetus as agent defaults to the social and discursive construction of the fetal subject*. Yet this need not be the case. Indeed, the mappings outlined earlier describe how the body mediates the social in that agency is enacted through it; as such, in ANT mappings the social need not be conceptualized as an entity or domain, nor as a product of magical thinking; rather, it can be conceptualized in terms of material forces that are affective and shared. Ironically, ANT mappings that disengage from the phenomenological experience of social energies flowing through the body invite interventions that deploy the very form of agency ANT seeks to disrupt: a humanist, discursive, and intentional agency that legitimates a rhetoric of individual bodily autonomy. Thus, for example, mappings that ignore the fetus's biochemical influence on the body, and the ways in which those biochemical and physiological influences are experienced, make possible political and legal provisions such as the Display Requirement, which seeks to "speak for" the fetus within a framework that casts it as an autonomous but silenced subject.

Agential Realism and the Display Requirement

In 1998, Karen Barad directly addressed how fetal ultrasound technologies interact with the fetus to produce an ontological legibility, specifically discussing Judith Butler's account of the role of ultrasound imaging in the interpellation of the subject. Barad points out the need to consider more seriously the lived reality of the body, and she does this by suggesting that Butler's performativity would be more useful if "iterative citationality" were recast as "iterative intra-activity" (p. 106). Barad's critique of the discursive here parallels ANT's critique of the social, in that discursive explanations can gloss over material interactions that are not fully recognized as agential in the network. However, because ANT does not permeate the lived body, the point Barad makes about the shortcomings of performativity-as-citational applies equally to Latour's performativity of identity groups (Strum & Latour, 1987). Although Latour does not specifically deal with fetal personhood, ANT serves as the predominant theoretical foundation for arguments of nonhuman agency and personhood; as such, it is important to note that women as a group have a great deal more to lose in terms of their own personhood, which is predicated on the concepts of individual agency and bodily autonomy that posthumanist theories erode.

Feminist materialisms are aligned with ANT in that they also emphasize the mutual affectability of agents. However, their careful consideration of the body prevents the maternal and fetal subjects from being placed in opposition to one another. In Barad's agential realism, biochemical changes in the pregnant body initiated by the fetus are themselves recognized as enactments of agency that do not require mediation by medical, legal, or state representatives; therefore, the idea of a fetal agency does not bleed into the idea of fetal subjectivity. As Barad (1998) explains,

> the fact that the fetus "kicks back," that there are fetal enactments, does not entail the concession of fetal subjectivity. Recall that the fetus is a complex material-discursive phenomenon that includes the pregnant woman in particular, in intra-action with other "apparatuses." And fetal enactments include the iterative intra-activity between the pregnant woman and her fetus. This formulation exposes the recently intensified discourse of hyper-maternal responsibility as a displacement of the real questions of accountability onto the pregnant woman who is actively constructed as a "mother" bearing full responsibility, and the full burden of accountability, for fetal well-being, including biological and social factors that may be beyond her control. (p. 116)

The "hyper-maternal responsibility" Barad mentions has only increased in its intensity since 1998, based on a number of converging factors that increase the social currency of the fetus. In many ways, the Display Requirement is a logical extension of the dynamic in which strangers feel the need to defend the fetus against the pregnant woman it inhabits. This sense that a pregnant woman is accountable both to the fetus and a general public that scrutinizes her choices on behalf of that fetus manifests itself with varying degrees of intensity and seriousness. For example, anyone who has been visibly pregnant is familiar with how strangers feel compelled to offer their input on whether one has a glass of wine, or is eating too much, or not enough, or is drinking coffee, ad infinitum. The over thirty-year reign of *What to Expect When You're Expecting*, the highly prescriptive "pregnancy Bible" that consistently tops the *New York Times* best seller list in its category, has also contributed to the culture of fetus as a social entity prior to birth.

While these everyday instances of the public policing of pregnant bodies do not necessarily seem insidious, they are part of the same overall orientation toward unreasonable maternal accountability that is specifically articulated through policies like the Display Requirement. At the extreme end of this dynamic is Purvi Patel's 2015 sentencing to twenty years in an Indiana prison for both feticide *and* child neglect after having a miscarriage (Chowdhury, 2015). The contradiction between the two charges is itself a demonstration of the ways in which early twenty-first-century policy on fetal personhood exists in a liminal space that can no longer be coherently expressed through rhetorics of bodily autonomy. Unfortunately, the ethical and legal complexities of Patel's case are too extensive to be fully addressed here; however, the example serves to demonstrate how, in issues of reproductive rights, pregnancy, and childbirth, rhetorics of bodily autonomy cannot avoid a conflict of interest between pregnant woman and fetus once the fetus has been constructed as a subject in its own right.

In contrast, and by virtue of its finely grained engagement with the body, Barad's agential realism casts the pregnant woman and fetus as an inseparable intra-active agential apparatus. The hormonal and biochemical changes that occur with a pregnancy are not, in this case, the ontologically distinct fetus acting on the body of the pregnant woman, but rather an extension of the biological phenomena of the reproductive cycle. In other words, the hormonal and physiological changes that occur after fertilization and implantation are not different in kind from the typical patterns of the menstrual cycle and the ways in which they influence affect, mood, and behavior. No two people will respond identically or even predictably to these changes, as their outcomes are subject to the convergence of additional material and social phenomena. The biochemical media of emotion are at once material, social, and discursive, in that an emotion is both the expression and result of a particular combination of chemical, physiological, and neurological responses. Each of these micro-phenomena brings agential force to bear on a pregnant woman's way of inhabiting or refusing to inhabit motherhood.

Some of these phenomena are initiated by the fetus as material agent. As the fetus develops, its presence triggers enormous surges in hormones at various points. Within days of fertilization, the implantation of the embryo elicits the development of the placenta, which is characterized as a "materno-fetal" organ and produces enough hormones to be considered a temporary endocrine organ. One of the earliest detectable changes is an enormous spike in human chorionic gonadotropin, followed by more gradual but sustained increases in progesterone, prolactin, and estradiol, all of which are associated with changes in affect and mood. In addition to the emotional manifestations of these shifts, there are increases in brain plasticity, allowing for changes in the "morphology, physiology, and function of many different brain structures," including the hippocampus and cortex, which are involved in maternal behaviors, learning, and memory, all of which can have behavioral implications (Workman, Barha, & Galea, 2011, p. 54). The physiological and biochemical changes can lead pregnant women to interact with their environments differently than women in the general population (Moya et al., 2014). One common way this manifests is through aversions to smells and foods, although many interactions are not detectable. Emergent behaviors form structural or biochemical changes; additionally, there might be socially inflected adaptive behaviors such as quitting smoking, reducing caffeine, or adopting a healthier diet (Moya et al., 2014). In addition to these bodily changes, both outside and within the conscious government of the pregnant woman, the discursive markers of pregnancy might also be performed, such as acquiring furniture, clothing, and other supplies for the baby's use upon arrival; referring to the fetus using gendered pronouns or a name; and sharing information and images (ultrasound and otherwise) of the pregnancy in person and via social media (Mitchell, 2001). From an agential realism perspective, motherhood *becomes* for the pregnant woman: each physiological and emotional response from the moment of conception is taken seriously as an agential element in the outcome of the pregnancy.

From an agential realism perspective, then, the ultrasound image is just one of many phenomena that will influence the pregnant woman's behavior and actions. Maternal accountability is not constructed just once, as a monolith, by the state; rather, it is a process of becoming, one in which atomic interactions are projected into affect and mood that are in turn projected into behavior. And, if we take seriously the affective and emotional phenomena emerging from physical changes, we must also take seriously the emotional and affective phenomena emerging from social context. The biochemical flood of fear and anxiety brings just as much to bear on subsequent action as does the biochemical flood of joy, both in the moment and over time (Damasio, 1994).

Because an agential realism perspective acknowledges the fetus's active biochemical and physiological role within maternal-fetal intra-active agency, it minimizes the possibility of conceptual bleed between the fetal agent and the fetal subject. The pregnant woman and fetus are a permeable entity ensconced in a moment that requires a "cut," which Barad (2007) defines as a "resolution of ontological indeterminacy . . . and *agential separability—the agentially enacted material condition of exteriority-within-phenomena*" (p. 175). In this case, ultrasound technology enacts the condition of exteriority by creating an observable exterior surface to the fetal body within the uterus—or, to return to Krøløkke's (2011) terms, it "out[s] the inner space" (p. 19). However, agential cuts are always made by unbounded apparatuses, and though we are ethically accountable for the cuts that we help make, we are never *solely* accountable. When the fetus is articulated as inert matter or a living object, a potential human or a legal person with constitutional rights, different apparatuses are making cuts to create the boundaries around those articulations. Ultimately, within Barad's framework, women are accountable not to the fetus as person nor to the state or public that would police her adopting the role of mother. Rather, those who would construct fetal personhood are accountable for the consequences of that articulation:

> The real questions of accountability include accountability for the consequences of the construction of fetal subjectivity, which emerges out of the particular material-discursive practices; accountability for the consequences of inadequate health care and nutrition apparatuses in their differential effect on particular pregnant women; accountability for the consequences of global neocolonialism, including the uneven distribution of wealth and poverty; and many other factors. (Barad, 2007, p. 218)

In other words, a woman who chooses to terminate a pregnancy is accountable for her role within an environment that is defined by a particular set of resources and circumstances.[18]

Barad's (1998) explication of accountability here emerges from her critique of Butler's (1993) version of matter, which the latter defines as *"a process of materialization that stabilizes over time to produce the effect of boundary, fixity, and surface"* (p. 9). Barad argues that these defining effects are in fact limitations of Butler's account of materialization as a whole. For the purposes of my argument, it is worth noting that, first, the limitation created by engaging with body only on the surface is one that applies equally to ANT. Second, but more importantly, a *fixed* materiality requires a categorical ontological status; when such an ontological status is, as Butler argues, a "dissimulated effect of power," the notion of basing ontological personhood on material constraints could be troubling. Barad's agential realism eschews the categorical enactment of power (as does ANT) in favor of the finely grained particularities of intra-agential action. In the context of fetal personhood, this means that the ontological personhood that would—if categorically state-mandated—serve as the basis for legal personhood remains primarily mediated by the pregnant woman. The woman is making an agential cut by choosing to define the ontological weight of the fetus in light of all of the circumstances surrounding the relationship. By this logic, a state that ascribes personhood to a fetus is accountable for providing the infrastructure of care for that fetus; in lieu of the ability to do so, the state must entrust individual women to determine the personhood of the fetus based on the medical, material, emotional, and environmental resources that will constitute the intra-active apparatus of life. The ethics are fluid, and the fetus, as part of the apparatus, participates in making the cut.

Accountability and Rhetoric

The interconnectivity and mutual affectability of agents in both ANT and agential realism flatten out relationships in ways that make apparent how agency is distributed and rarely located within a single bounded entity. Actions are never fully autonomous. What Barad offers that Latour does not, however, is a converse sense of distributed accountability.

ANT's version of distributed accountability is a negative one: individuals within a network may not intend or even be aware of the effects of the larger network to which their actions contribute. In other words, as agency is distributed, actants are released from accountability. Barad, however, offers a notion of accountability that is lived. For example, as quoted earlier, Monica Casper (1998) can draw a hard line between the personhood of her cats and a fetus because the extension of personhood to a cat does not impinge on her own personhood as a woman, but also because she is referring to fetuses as an abstract category of being, while granting personhood to her particular cats. Accountability is specific and is generated through shared emotional and affective resonances—while one might not be comfortable granting personhood to fetuses as a category, one might want to grant personhood to one's own fetus. Agential realism allows for personhood to emerge when the necessary conditions are present; those conditions are less about characteristics attributed to the fetus (or other nonhuman person) and more about creating an environment that can sustain personhood.

Ironically, though the category of "mother" requires moral accountability for factors well beyond a single person's control, the rhetoric of "acceptable abortion" requires that women be cast as at the mercy of circumstances beyond their control (Settles & Furgerson, 2015). This is clear in the frequent pro-life provisions for rape, incest, or danger to the life of the mother, but it manifests in more subtle ways as well. For example, Paige Settles and Jessica Furgerson's (2015) analysis of abortion narratives suggests that abortion is only deemed acceptable as a last resort, and, further, "public support is given only to women who successfully defend their right to abort" (p. 16). In Barad's terms, the agential cut must be rendered as necessary by extreme mitigating factors such as a threat to life of mother; the decision must be ratified by a demonstration of suffering, remorse, or redemption. The woman is absolved for her enactment of agency only by distancing herself from it.

At the same time, taking action based on agential accountability for material conditions is not considered an "acceptable" abortion. There is resistance to the idea that material supports should influence a decision about terminating a pregnancy, partially because of the desire for a categorical personhood. This is in spite of the fact that the vast majority of women who terminate pregnancies do so for reasons of material support. Seventy-four percent of women surveyed said that having a child would interfere with "education, work, or ability to care for dependents"; 73 percent said they could not afford a baby, as opposed to 32 percent who cited more abstract reasons such as not being ready to have a child (Finer, Frohwirth, Dauphinee, Singh, & Moore, 2005, p. 110). In these cases, women opt not to deploy the networks that construct fetal subjectivity, based on their assessment of the conditions of possibility for personhood. The appeal to privacy that served as the basis for *Roe v. Wade* maintains the necessary space of becoming between ontological perceptibility, motherhood, and legal personhood.

In this way, the issue of abortion becomes less an argument about individual freedom and bodily autonomy and more about social and public apparatuses that support motherhood, not as a category but as a practice. Prenatal and pediatric care, nutrition and education, parental leave and child care, resources for shelter and daily necessities, and support networks all become the conditions of possibility for fetal personhood. While ANT would acknowledge that all these things matter in raising children, agential realism's attention to the body and to lived experience moves us from the notion of tracing patterns of influence in distributed agency to recognizing that an ethic of distributed agency requires an ethic of distributed accountability, where we all bear some responsibility for the conditions of those whose lives we share. While *"neither distance nor empathy* defines well-articulated science" (Latour, 2004, p. 219), empathy, emotion, and attention to the conditions of everyday life are central to feminist rhetorics and to a feminist politics. In part because of its explicit feminist intention, but primarily because of the seriousness with which it regards lived experience, Barad's theory of agential realism allows for an ethics of inclusivity that broadens the category of personhood to include entities with which there is an affective resonance or other empathic connection, while still dissolving the boundaries between nature, culture, science, material, discourse, and so on. The material and the social adhere to one another in the micro-interactions that accumulate through the embodied experience of interacting with the environment: agential force comes from the aggregation of chemical, neural, and electrical responses that make up human connection and manifest as social behaviors. The inclusion of bodies as vital, dynamic, and ontologically rich ecologies both in themselves and in conjunction with the environment allows for the reciprocity and sense of humility that are the most powerful aspects of flat ontologies.

Notes

1. In fact, the idea of a woman "having more" agency than a fetus is one of the reasons fetal personhood was rejected by counsel on both sides of the argument in *Roe v. Wade* and again in 1992 in *Planned Parenthood v. Casey* (Condit, 1990; Gibson, 2008; Langford, 2015). Celeste Condit (1990) argues that three major points emerged from *Roe v. Wade*'s oral arguments:

> (1) that the only thing that could possibly outweigh a woman's Right to Choice would be another person with an independent and more fundamental Right to Life, (2) that the status of the fetus was clearly a constitutional issue which the Court had the *necessity* of deciding, and (3) that to label a fetus a person entailed legal and logical inconsistencies the law could not tolerate, *absent a well-argued reason external to the law for doing so* [emphasis added]. (p. 109)

The spate of fetal personhood bills that have been proposed and passed by state legislatures since 2008 suggests that just such a reason external to the law has emerged: fetal ultrasound technologies.

2. A distributed agency is one in which persons, animals, machines, technologies, and environments interact to create particular effects. No single entity within the network or assemblage owns the effects; rather, each human and nonhuman entity that contributed to an outcome has enacted agency within that network. So, for example, when a surgeon performs a lifesaving heart transplant, that surgeon is not the sole agent of the action; rather, the surgeon, each member of the surgical team, the patient, the donor, the hospital itself, the helicopter that transported the donated heart, the helicopter pilot, all the technologies associated with diagnosis, the surgery itself, and the recovery, anesthesia, painkillers, and immunosuppressants, and so on, are all considered agents that contributed to the effect of saving a life. While agents such as the chemicals used in anesthesia, which have effects that cause changes in the body of the patient, can be said to have agential force, they cannot be said to have subjectivity or intention.

3. When dealing directly with Latour, I use "actants"; otherwise I will use "agents" to refer to both human and nonhuman entities that affect other human and nonhuman entities.

4. The ultrasound must be performed and described between seventy-two and four hours prior to the procedure.

5. Georgia, Indiana, Kansas, Michigan, Missouri, Nebraska, North Carolina, Oklahoma, South Carolina, Utah, Virginia, and Wisconsin.

6. Arkansas, Georgia, Idaho, Michigan, Nebraska, Ohio, South Carolina, Utah, and West Virginia.

7. Alabama, Arizona, Florida, Indiana, Kansas, Mississippi, North Carolina, Ohio, Oklahoma, and Virginia.

8. Louisiana, Texas, and Wisconsin, though these are currently unenforceable pending court rulings on constitutionality. The North Carolina provision, as well as one in Oklahoma, were implemented through executive order.

9. In these cases, sex is identified as binary and based on genital appearance.

10. According to the Centers for Disease Control and Prevention's 2014 report, the number was 91.4 percent in 2011. According to the Guttmacher Institute, it is 89 percent.

11. This is even earlier than a fetal Doppler can pick up an audible heartbeat, usually at eight or nine weeks.

12. Asymmetries in the context of efforts to legalize the personhood of fetus are not leveled by *Roe v. Wade* and subsequent protections to reproductive freedom because they do not typically engage with the question of fetal personhood; rather, they are based on the right to privacy. The right to privacy, while not explicitly granted in the Constitution, has a number of precedents dating from 1891 in which the court guaranteed a right of personal privacy. Different courts have located the rationale for such a right in several locations in the Constitution (*Roe v. Wade*). In the case of *Roe v. Wade*, the right of privacy was located in section 1 of the Fourteenth Amendment, which restricts state action on citizens.

13. Limiting the number of diagnostic articulations limits the number and location of choices to be made. Importantly, the moment of accountability comes in the political decision of whether and how to treat a patient—a decision that identifies the location of choice as the choice is made. Although in theory that choice should be made by the patient, the performance of patient also serves to effect the site of choice. Just as a method of diagnosis creates a reality in which a patient either does or does not "have" a condition or disease, so too does the decision of whether and how to treat the diagnosis creates a patient (Mol, 2002, p. 86). For example, a patient-customer is "making choices about discrete goods available on a market," whereas a patient-citizen is "trying to organize the healthcare system for the benefit of all" (p. 86). Though the U.S. health-care system generally casts patients as customers, the Display Requirement clearly enacts patient-citizens.

14. From a medical perspective, a "shared, coherent ontology is not required for treatment and prevention practices. Incompatibilities between objects enacted are no obstacle to medicine's capabilities to intervene—as long as the incompatible variants of an object are separated out" (Mol, 2002, p. 115). In other

words, the necessity of choice implied by an ontological politics is suspended in a medical context; the site of choice need not be consistent (Mol, 1999, p. 86). In the context of fetal ultrasound, its use as a diagnostic tool does not displace the pregnant woman as mediator of the ontological status of the fetus as person. The open-endedness of the ontological question from a medical perspective allows the pregnant woman to remain the site of choice; it is the legal mandate of a diagnostic procedure that displaces her, moving the site of choice to the state through enacting the social category of mother.

15. For example, most of the women cited by Condit (1990) made statements such as: "I really feel that the thing that makes it real is the choice to *have* the child" (p. 183).

16. While this statement carried some fairly radical implications in 1994, in 2014 Marlise Muñoz's husband had to fight to remove her from life support after her brain death, as it is against Texas law to remove a pregnant woman from life support. Muñoz was fourteen weeks pregnant when she was declared brain-dead (Shoichet, 2014).

17. Charlotte Kroløkke's (2011) study, which examines descriptions of fetal behavior by ultrasound technicians and parents, affirms Butler's argument about the interpellation of the fetal subject via gendering. Specifically, Kroløkke found that the same actions were described differently depending on the sex of the fetus (much as they are for babies and children). Crossed legs, which make it difficult to determine genital sex, are described as modesty in girls and stubbornness in boys; kicking is described as dancing in girls and soccer playing or running in boys, and so on.

18. This contrasts with Mol's (1999) accountability, since the occasion for choice is not a site among a network of human and nonhuman sites; rather, it is a sufficiently realized set of conditions of possibility.

References

Barad, K. (1998). Getting real: Technoscientific practices and the materialization of reality. *Differences, 10*(2), 87–128.

Barad, K. (2007). *Meeting the universe halfway: Quantum physics and the entanglement of matter and meaning.* Durham, NC: Duke University Press.

Braidotti, R. (2006). Posthuman, all too human: Towards a new process ontology. *Theory, Culture, and Society, 23*(7–8), 197–208.

Butler, J. (1993). *Bodies that matter: On the discursive limits of "sex."* New York, NY: Routledge.

Casper, M. J. (1994). Reframing and grounding nonhuman agency: What makes a fetus an agent? *American Behavioral Scientist, 37*(6), 839–856.

Casper, M. J. (1998). *The making of the unborn patient: A social anatomy of fetal surgery.* New Brunswick, NJ: Rutgers University Press.

Centers for Disease Control and Prevention. (2014, November 18). Reproductive health: data and statistics. Accessed 20 December 2014. Retrieved from http://cdc.gov/reproductivehealth/data_stats/

Chowdhury, J. (2015, March 31). Indiana sentences Purvi Patel to 20 years for feticide. *NBC News.* Accessed 25 May 2015. Retrieved from http://www.nbcnews.com/news/asian-america/indiana-has-now-charged-two-asian-american-women-feticide-n332761

Condit, C. M. (1990). *Decoding abortion rhetoric: Communicating social change.* Urbana: University of Illinois Press.

Damasio, A. R. (1994). *Descartes' error: Emotion, reason, and the human brain.* London, UK: Penguin Books.

Finer, L. B., Frohwirth, L. F., Dauphinee, L. A., Singh, S., and Moore, A. M. (2005). Reasons U.S. women have abortions: Quantitative and qualitative perspectives. *Perspectives on Sexual and Reproductive Health, 37*(3), 110–118.

Gibson, K. L. (2008). The rhetoric of *Roe v. Wade*: When the (male) doctor knows best. *Southern Communication Journal, 73*(4), 312–331.

Guttmacher Institute. (2014, December 1). State policies in brief: Requirements for ultrasound. Accessed 18 December 2014. Retrieved from http://guttmacher.org/statecenter/spibs/spib_RFU.pdf

Haraway, D. J. (1991). *Simians, cyborgs, and women: The reinvention of nature.* New York, NY: Routledge.

Haraway, D. J. (1997). *Modest_witness@second_millenium. FemaleMan© _meets_OncoMouse™: Feminism and technoscience.* New York, NY: Routledge.

Hekman, S. (2008). Constructing the ballast: An ontology for feminism. In S. Alaimo & S. Hekman (Eds.), *Material feminisms* (pp. 85–118). Bloomington: Indiana University Press.

Kroløkke, C. H. (2011). Biotourist performances: Doing parenting during the ultrasound. *Text and Performance Quarterly, 31*(1), 15–36.

Langford, C. L. (2015). On making <person>s: Ideographs of legal <person>-hood. *Argumentation and Advocacy, 52*(2), 125–140.

Latour, B. (2004). How to talk about the body? The normative dimension of science studies. *Body and Society, 10*(2–3), 205–229.

Latour, B. (2005). *Reassembling the social: An introduction to Actor-Network -Theory.* New York, NY: Oxford University Press.

Mitchell, L. M. (2001). *Baby's first picture: Ultrasound and the politics of fetal subjects.* Toronto, Canada: University of Toronto Press.

Mol, A. (1999). Ontological politics: A word and some questions. *Sociological Review, 47*(S1), 74–89.

Mol, A. (2002). *The body multiple: Ontology in medical practice*. Durham, NC: Duke University Press.

Moya, J., et al. (2014). A review of physiological and behavioral changes during pregnancy and lactation: Potential exposure factors and data gaps. *Journal of Exposure Science and Environmental Epidemiology, 24*(5), 449–458.

National Institutes of Health. (2003, December). A thin blue line: The history of the pregnancy test kit. Accessed 15 October 2016. Retrieved from https://history.nih.gov/exhibits/thinblueline/index.html

Planned Parenthood v. Casey, 505 U.S. 833 (1992).

Roe v. Wade, 410 U.S. 113 (1973).

Sanger, C. (2008). Seeing and believing: Mandatory ultrasound and the path to a protected choice. *UCLA Law Review, 56*, 351–407.

Settles, P., & Furgerson, J. (2015). The acceptable abortion: Thematic consistencies of prominent narratives within the U.S. abortion debate. *Kentucky Journal of Communication, 34*(2), 16–39.

Shoichet, C. E. (2014, January 30). Husband of brain-dead Texas woman: "I asked God to take me instead." *CNN*. Accessed 25 May 2015. Retrieved from http://www.cnn.com/2014/01/29/health/texas-pregnant-brain-dead-woman/index.html

Strathern, Marilyn. (1996). Cutting the network. *Journal of the Royal Anthropological Institute, 2*(3), 517–535.

Strum, S. S., & Latour, B. (1987). Redefining the social link: From baboons to humans. *Social Science Information, 26*(4), 783–802.

Stuart v. Camnitz, 14-1150 U.S. 4th Circuit (2014).

Sturman, S. (2006). On black-boxing gender: Some social questions for Bruno Latour. *Social Epistemology, 20*(2), 181–184.

Walsh, J., Hepper, E. G., & Marshall, B. J. (2014). Investigating attachment, caregiving, and mental health: A model of maternal-fetal relationships. *BMC Pregnancy and Childbirth, 14*, doi: 10.1186/s12884-014-0383-1. 1–9.

Workman, J. L., Barha, C. K., & Galea, L. A. M. (2011, October 3). Endocrine substrates of cognitive and affective changes during pregnancy and postpartum. *Behavioral Neuroscience*. Advance online publication. doi: 10.1037/a0025538. 54–72.

3.
"The Inconvenience of Meeting You"

Rereading Non/Compliance, Enabling Care

Catherine Gouge

Uncooperative patients are often cited as a problem for modern medicine. "There are certain characteristics I've noticed that my medical teams like," one fourth-year medical student writes of her experiences "rounding" on patients:

> A "good attitude," politeness, agreeability, compliance with treatment, and the ability to understand without asking too many questions. Fall down on one or more of these and during a busy morning we may round on you last, or we won't round on you as a team at all. . . . so as to spare the members of the team not directly involved in your care the inconvenience of meeting you. (Yurkiewicz, 2013)

This account reflects the conventional and persistent understanding of noncompliance among many health-care professionals, which holds that uncooperative patients are "inconvenient" at best and either intentional or accidental villains, at worst, in their own care. In fact, patient "noncompliance with advice or treatment recommendations" was the top complaint of doctors in the 2011 *Consumer Reports* survey "What Doctors Wish Their Patients Knew," which notes that such behavior "affect[s] their ability to provide optimal care." Patient nonadherence with treatment recommendations has been said, furthermore, to waste $289 billion dollars annually in the United States in disregarded or misallocated health resources (Fung, 2012).

Inside of a framework that understands patient divergence from medication or treatment recommendations as an obstacle to care, part of the role of health-care professionals is to manage and sometimes mediate individuals' self-destructive tendencies so that they can be saved from themselves.[1] An Aaron Bacall cartoon (2011) plays to this understanding of the "problem of noncompliance" in its depiction of one man in a lab coat saying to another man in a lab coat, "My diabetic research shows that test subjects are 98% more likely to take their diabetic pills if the pills are covered in chocolate." In spite of the fact that a person with diabetes might need to perform over three thousand behaviors a year to be considered perfectly compliant (Ofri, 2012), the "joke" in this turns on the notion that the challenges posed by so many noncompliant, self-destructive diabetics necessitate such measures in order for doctors to protect people with diabetes from themselves—a notion disability studies has critiqued for the ways that it reinforces the association of normalcy and health with the "exercise of power and control" (Hall, 2011, p. 5).

Biomedical literature, "as a whole," Judy Segal (2005) writes, "fails to acknowledge that the medical model that produced the problem of noncompliance is limited in its ability to solve it" (p. 137). "Noncompliance," she argues,

> does not need to be obsessively measured, and it does not, as a concept, need to be rejigged out of existence. The actions of patients need to be studied not mechanistically but with an appropriately complex theory of human persuasion and human judgment: a theory of rhetoric. (p. 151)

Indeed, many have identified the so-called problem of noncompliance as one precipitated and exacerbated by issues that, whether they are treated as such or not, are distinctly rhetorical: a failure to persuade and motivate the called-for, expected, or agreed-on response (Schwartz, 2011), as well as the ambivalence of both patients and physicians about expertise that stems from the "equality rhetoric" (Segal, 2005) of concordance models, the purpose of which is to align therapeutic goals with the goals of those seeking care. Additionally, trust is often cited as an issue: clinicians don't trust patients to comply, and patients don't trust clinicians to understand all that they are dealing with in their experiences of illness. And following from that, the ethos of both care providers and patients in the clinical encounter is compromised (Mayo Clinic, 2010). The trust-ethos dynamic, as those accounts describe it, is endlessly recursive and undermines the "therapeutic alliance" (Behforouz, Drain, & Rhatigan, 2014) because patients often do not find the expertise of the physician, the clinical encounter itself, and, perhaps, the experience of having tried to follow the advice of the physician to be persuasive and motivating enough to act in accordance with it. Lack of trust begets noncompliance; noncompliance begets lack of trust; and so on.[2] Rarely, though, does noncompliance as an issue that rhetoricians might help address move beyond a discussion of expertise, the delivery of information, and the persuasiveness of arguments made by health-care participants.

In fact, though the patient-centered care movement has been working for decades to transform both the power dynamics and nomenclature of therapeutic relationships (from noncompliance to nonadherence to concordance), compliance expectations are still very much a part of biomedical culture, and a new set of compliance expectations for people who are believed to be at risk of life-threatening and chronic illnesses has taken shape. In response to this, informed by scholarship in rhetoric, feminist and disability studies, medical anthropology, and new materialist feminisms in science studies, this chapter proposes that if we are going to design more ethical, effective, and efficient systems of care, we must acknowledge and address the assumptions that continue to enable a culture of compliance in biomedicine, and we must challenge the disability system to which they are bound.[3] Because compliance expectations have historically been premised on very narrow understandings of agency and identity, and because the primary counterstrategy by many responding to the issue has been to argue against the attention paid to noncompliant or divergent behaviors, efforts to radically transform the expectations of health-care participants (both patients and providers) have been limited. Through an extended theoretical engagement with the "problem of non/compliance" and a discussion of several examples, this chapter argues that we must pay attention not only to compliance assumptions but also to behaviors considered to be noncompliant. In so doing, we become more aware of how divergent acts productively challenge ideas about what "care" is and should be, and better able to develop useful and just models for understanding divergent behavior in health and medicine.

In making this argument, I follow many disability scholars who question the deficit-model assumption that medical treatment and care are those things that support progress along an ideal wellness path and that divergence from such paths ought to be fixed via education, persuasion, or correction. The purpose of this extended engagement is to retheorize care as a process that ought to make sense of noncompliance as a rhetorically productive phenomenon and that ought, therefore, to value listening more closely to divergent practices in health and medicine for what they can teach us—about compliance frameworks; the needs, goals, and means of health-care participants; and complex ecologies of care. As an extension of this line of inquiry, I propose an approach, which might be described as a "kairology of care," that foregrounds the situated nature of the experience of illness and the value of attunement to material-discursive contexts and influences.

The argument in this chapter is predicated on a belief in the ontological inseparability of material-discursive practices in and contexts for care and, because of this, has important implications for feminist rhetorical science studies. First, this chapter responds to recent calls from feminist scholars that we question tendencies to privilege "the discursive" and thus to limit feminist analyses of biology to only its critique.[4] Second, because the argument in this chapter draws on new materialist feminist work in science studies to explore the rhetorical nature of behaviors in health and medicine, it offers a way of thinking about bodily rhetorics that extends beyond conventional humanist accounts of persuasiveness and influence and, in so doing, adds to prior work on embodied rhetorics.[5] Finally, this chapter responds to Jay Dolmage and Cynthia Lewiecki-Wilson's (2010) call for scholars in feminist rhetorics to explore the "generative potential of an alliance between disability studies and feminisms" (p. 24) as we think through the challenging issues associated with bodily rhetorics.[6] Ultimately, because this chapter calls into question assumptions about the ethical behavior of health-care participants, the shame and blame reactions a culture of compliance tends to summon, and the humanist ideas of agentic action on which compliance expectations are predicated, my hope is that it can support a revaluing of noncompliance that will be useful to understanding behaviors in health and medicine and that will work with other intersectional projects to "read with and against the practices that produce normalizing categories of all kinds" (p. 38).[7]

The Non/Compliance Problem

As a master narrative of normalcy and control in biomedicine, the compliance mandate and a fixation on securing it continues to influence clinical encounters, treatment protocol, and research studies. Compliance metrics still broadly inform, for example, the strategies practitioners use to communicate with patients: communication that leads to compliance is often labeled "effective," and that which does not lead to compliance is often labeled "ineffective." Using this logic, clinical studies attempting to address the problem of improving patient discharge communication have, over the last several decades, measured the impact of specific communication strategies on patient compliance with prescribed treatments (Agency, 2012; Gottlieb, 2000; Showalter, Rafferty, Swallow, Dasilva, & Chuang, 2011).

This logic of compliance is manifested in at least two different sets of compliance expectations in modern medicine (what I refer to as Compliance 1.0 and 2.0 later on), and both do the following: treat "health" as a clearly identifiable destination (in spite of the fact that it has historically been notoriously difficult to define); assume that there are ideal, linear, disease-specific paths that can take us there; and understand "care" in terms of identifying the most appropriate path, choosing it, and then following it. Inside such a logic of compliance, "progress" is thus defined as forward movement along prescribed paths. Recognizing these sets of compliance expectations, which work together in modern medicine to reinforce assumptions about agency and hopes of health-care participants, can help us understand how the prominence of compliance as an ideal, value, and privileged role has undermined biomedicine's ability to engage productively with the divergent behaviors of health-care participants and, thus, how it has disabled many practices and processes in health and medicine meant to support care. Furthermore, challenging assumptions about agency that underlie these expectations makes visible one of the critical weaknesses of the culture of evidence-based medicine: that judgment is often premised on rigid, dichotomous distinctions between the material/biological and the discursive/psychological. It also discloses the ways that biomedicine often identifies the locus of agency and, therefore, the ethical responsibility to comply within autonomous individuals[8]—often minimizing or, as David Martins (2005) describes it, "impoverishing" context.[9] Because of this, identifying the structure of compliance expectations is both essential to retheorizing noncompliance and to understanding why paying attention to divergent behaviors is critical to improving systems of care.

Compliance 1.0

Technologies of access to the anatomical body and other medical advances over the last couple centuries have contributed to broad acceptance of the concept of a standardized body in Western biomedical culture. The version of compliance in modern medicine that most of us are familiar with is predicated on ideas about the value of this standardized model of the body and, therefore, to ideas about biological normalcy and deviance. In modern medicine, the way of knowing the body from which this Compliance 1.0 framework derives began to gain traction roughly around the time that the anatomized body, made possible through dissection, became more accessible for medical study—during the emergence of what Michel Foucault (1973) and others before him have called the "anatomo-clinical method" (p. 4). As dissection started gaining prominence as a privileged method of learning about the body, so too did an idea of a universal human biology and the idea of "the standardized body." Medical anthropologists Margaret Lock and Vinh-Kim Nguyen (2010) write that, at this time, "individual bodies were 'normalized' both biologically and statistically" and "bodily variation began to be defined in terms of deviation from a statistical norm or average" (p. 32). The most familiar example of this for most people is the "routine blood test, the results of which are interpreted according to 'normal values' that distinguish 'healthy bodies' from the pathological" (p. 32). Since then, the idea of a universal human biology that extends from the standardized body and its attendant statistical norms has been used to identify and manage disease across cultural, geographical, and individual boundaries. And the results of this have not been entirely positive.

According to Foucault (1973), the anatomo-clinical method that characterizes biomedical practice is a particular method of medical perception grounded in a notion of "mechanical objectivity" (p. 108). And this way of seeing and knowing the experience of illness has contributed to the expectation that we assume what functional sociologist Talcott Parsons first identified in 1951 as the "sick role," a role considered to come with both rights and obligations. The "rights," according to Parsons, include the "exemption from normal responsibilities" and the "right to be cared for," while the "obligations" include the "obligation to want to become well" ("being ill" must be seen as undesirable) and the obligation to seek and cooperate with treatment (the common goal of "getting well" is mandatory) (Lubkin & Larsen, 2006, p. 25). The sick role, as many critical medical studies scholars have noted, "require[s] an absolute surrender to the care of a physician and the institutional structures of medicine" (Diedrich, 2007, p. 3). This notion that patients occupying the sick role ought to surrender completely to the judgment of the physician has, of course, led to many oppressive attitudes, beliefs, and practices.[10]

In this version of the compliance mandate, the locus of agency for those who assume the sick role is *reactive*. It is located in a person's willingness and ability to accept his or her diagnosis and comply with the treatment the doctor deems best. The Compliance 1.0 paradigm, therefore, assumes that patients are autonomous subjects making choices and taking an "active role" in their own care (though what can count as active is defined by others), and noncompliance, defined in opposition to this, is considered to be passive and self-destructive, a failure to make the right choices about care and cope with treatment. Those who are not compliant are thought to be interfering with treatment, at best, and posing dangerous (even malicious) public health risks at worst.

Compliance 2.0

Compliance 2.0 has its beginnings in what Arthur Frank (1991) and others have recognized as the "overturning moment" of modern medicine—the moment when we became a "remission society" in which chronic illness is a growing concern and the "boundaries between health and illness are permanently disrupted, thereby challenging the dichotomous formulation of health as the norm and illness as that which deviates from the norm" (Diedrich, 2007, pp. 2–3). As we have transitioned into a "remission society" (Frank, 1991), the dominant identity for people when they are patients has shifted, as many medical sociologists and anthropologists have recognized, from the "sick role" to the "at-risk" role. The at-risk role refers, as it sounds, to a state of being that exists before the onset of illness, during which we are at risk for things like type 2 diabetes, heart disease, high blood pressure, or cancer. Indeed, genomic research—which promises to personalize medicine as a corrective to the "one dose fits all" of "population" medicine—also promises to document our entrance into a "mutational community" of at-risk others before we are even born (Topol, 2012, p. 25). According to the at-risk role as it is taking shape, we are all at risk of something. And, in this way, we are all always and already medicalized. The version of compliance that aligns with this at-risk role accordingly requires that we consent to our construction as overdetermined, medicalized subjects—even before we are ill. In this version of compliance, anyone who fails to be proactive, anyone who fails to follow the "best practices" to prevent the onset of illness, is implicated. Those implicated by the Compliance 2.0 paradigm are often accused of being negligent, passive about their own care, and weak willed.

Kelly Happe's compelling book *The Material Gene: Gender, Race, and Heredity after the Human Genome Project* (2013) addresses the ways that genomic research is "defining rather than revealing" (p. 147) categories of at-risk people and so addresses a critical Compliance 2.0–related issue. Those affected will be expected to proactively address their at-risk status or chance being considered irresponsible and noncompliant—like those referred to as "hypersusceptible" (p. 142), for example, categories that are, Happe notes, "dependent on normative conceptions of the gendered, raced, and bounded body" (p. 3). *The Material Gene* begins with an account of a woman, "A.H.," whose experience is influenced by both Compliance 1.0 and 2.0 expectations. A.H. underwent several procedures in 1999 to remove her breasts, ovaries, and uterus. These measures, Happe notes, were taken in the hopes that they would extend the woman's life by about four to six years, but they were measures that were decided on in spite of the fact that there was no evidence of illness or disease present in her body. A.H. was not sick, and she did not have cancer. But she was "at risk" because she had tested positive for the BRCA gene mutation.

Though Happe doesn't frame it this way, A.H.'s treatment was deeply influenced by a newly forming set of compliance expectations for bodies at risk—Compliance 2.0—but the decision to proceed with such invasively preemptive measures was also dependent on 1.0 fantasies that patients in recovery (sick bodies asking for support and, therefore, presumably pursuing wellness) are responsible for following and tolerating prescribed treatments. The 2.0 medicalization of bodies at risk has, in this way, contributed to what medical anthropologists note is a frequent consequence of medicalization: the "masking [of] significant historical, social, and political variables that contribute to illness and distress" (Lock & Nguyen, 2010, p.78). Indeed, according to Happe, "the decision model [A.H.'s doctors] employ assumes that there is a complete mastectomy and no reconstruction (which in fact A.H. had), and that the patient *tolerates*, and *complies* with treatment during many years of potentially dangerous hormone replacement therapy" (2013, p. 1, emphasis mine).

A.H.'s story illustrates the ways that genomics, as Happe argues, "epistemologically and rhetorically . . . effects the dematerialization of the body and its embeddedness in historically specific environments when biological matter is translated into the language of gene sequences" (p. 2). The case of A.H. also demonstrates, I would add, how genomics—as a way of knowing the body—supports Compliance 2.0 expectations with regard to health decisions, naturalizing some choices as "right" and ethical and privileging certain ways of knowing over others. Indeed, genomics perpetuates a view of patient compliance that tends to ignore the lives of patients beyond their status as "sick" or "at risk" and, in so doing, supports an oversimplified view of ethical decision-making for health-care participants. Supported by Compliance 1.0 fantasies that people, when they are recovering, will find a way to endure treatment in a kind of Cartesian mind-over-matter triumph of will, genomics encourages Compliance 2.0 assumptions about agency and ethical decision-making (act preemptively to reduce risk). In so doing, it circumscribes expensive and risky preemptive action as the right, compliant thing to do. As Stuart Murray and Deborah Steinberg (2015) argue, this is one of the "paradoxical effect[s] of neoliberal biopolitics": "A heightened sense of individual responsibility—but one that is thoroughly mediatized and regulated, producing the illusion of freedom and democracy, while circumscribing and policing these terms ruthlessly" (p. 131).

Indeed, in a culture of compliance that aligns compliant behaviors with "normal" and "right," images of the bodies of those who have been noncompliant are mobilized as warnings—"inferior," "grotesque," and "ugly," "spectacles of otherness" (Garland-Thomson, 1997, p. 8). The thing is, as Lennard Davis writes, "under normalcy, no one is or can be normal" (2013, p. 117). And yet the bodies of so-called abnormal, deviant patients are strategically rematerialized as cautionary tales, warnings not to be "bad"—or else. As Garland-Thomson (1997) has argued of disability in an "economy of normalcy," the noncompliant body—often dematerialized in measurement studies—is rematerialized strategically to shore up the boundaries of compliance and to further associate compliance with health and ableism as a way of motivating people to follow the advice of their physicians.[11] Indeed, some startling persuasive techniques have been proposed for physicians to get patients to "act better" (Schwartz, 2011).

An article on the website *Podiatry Today*, for example, advises physicians that if they want to motivate people with diabetes to comply with treatment, "nothing hits a patient between the eyes like a photo of amputated toes or a partial foot amputation" (Hall, 2006). In this way, the rematerialization of noncompliance as "disfigured" body parts to secure compliance supports a logic of compliance that marks noncompliance as self-destructive and deviant. The abject figure of "the noncompliant patient" also helps define what constitutes ideal, compliant patient subjectivity—a model stemming from a fraught and exclusionary history of medicalizing deviance. "Outlined by an array of deviant others whose marked bodies shore up the norm's boundaries" (Garland-Thomson, 1997, p. 8), the "compliant patient" often circulates as an abstraction in modern medicine—an "unmarked disembodied normate" (p. 136) bolstered by a disability system that "excludes the kinds of bodily forms, functions, impairments, changes, or ambiguities that call into question our cultural fantasy of the body as a neutral, compliant instrument of some transcendent will" (2011, p. 17).

The Compliance Mandate

As many disability scholars and medical anthropologists have noted, the development of population medicine and its associated statistical norms put doctors in the role of regulating the "rule of normalcy" (Conrad & Schneider, 1992; Davis, 2002). "Under the rule of normalcy," as Davis (2002) explains it, "the physician becomes instrumental in determining whether each citizen is 'normal' or 'abnormal'" (p. 115). Like other conceptions of "normal" (Davis, 2013; Garland-Thomson, 2011; Siebers, 2008), compliance expectations both mark and police narrow assumptions about agency and identity for people when they are patients. They circulate as a corollary to what Jay Dolmage (2014) argues resulted from the "commingling" of conceptions of pre-nineteenth-century valuing of a bodily "ideal" and post-nineteenth-century ideas of "normal": a "normative mandate—both to uphold the fiction of perfection and to generate the systematic self- and other-surveillance and bodily discipline of normative processes" (p. 23). Indeed, it is only when viewed through and against compliance frameworks that divergent behaviors in health and medicine, like divergent biology, become deviant.[12] Illustrating the ontological inseparability of "the material" and "the discursive," broad investment in the value of a standardized body to biomedicine has contributed to a repressive discourse of "normal" and a corresponding compliance mandate that tends to identify difference as either evidence of disease or deviance. In this way, while providing a useful, generalizable object of study and a seemingly progressive argument for treating everyone everywhere the same, standardizing human biology has also drawn attention away from the specific material conditions of and even biological influences on patient experiences of illness and, in so doing, has contributed to a narrow understanding of "individual agency and responsibility [that] disguises the social and political origins of disease and illness" (Lock & Nguyen, 2010, p. 79).

As a result, the privileged position of compliance—like whiteness, the able body, and masculinity—circulates mostly unexamined in Western biomedicine as the "unmarked normal" and is presumed to be the responsible, ethical, and capable condition of ideal patients. This idealization is a fantasy of patients who are conduits for the healing powers of medicine and do not interfere with medicine's ability to provide treatment. No one who seeks care during their lifetime is perfectly compliant, however, and health-care professionals know this. Noncompliance with medication and behavioral prescriptions is quite common, and compliance rates tend to trend down over time, making noncompliance arguably more "the norm" than the exception (Agency, 2012). And yet noncompliance is routinely cast as a deviant subject position and stigmatized as the purview of bad patients who fail to manage their own care appropriately. As one physician with multiple sclerosis writes of her experience,

> I am deliberately choosing to ignore my doctors [sic] advice. He wants me to be on medication, but I quit a month ago. While I trust that I am doing the right thing, I can't help but feel that I am a "bad" patient. (Zimmerman, 2012)

Though Compliance 1.0 and 2.0 frameworks for making sense of patient behavior are differently scaffolded and offer different versions of what it means to be a "good" patient—accepting the "sick" or "at risk" role and being reactive and proactive—these frameworks have in common that following them is understood to be the right choice, an ethical imperative, and evidence of successful coping. That is, if we are ethical, compliant patients, we decide what is right—following the advice of our physician—and choose to act accordingly.

Notably, agency in both 1.0 and 2.0 paradigms is often understood to be asynchronous with illness—framed as either prior to or after the onset. This is, of course, quite different from the actual experience of illness during which symptoms and effects come and go, are variable, and so are sensed and addressed repeatedly with whatever resources are available in that instant, as they are disclosed, moment by moment. Though Compliance 1.0 and 2.0 offer two versions of asynchronous interventions, associated with different, historically specific, privileged ways of knowing the body, they often work together in modern medicine in a way that perpetuates narrowly defined, privileged assumptions about agency and identity that ignore the specificity of our relationship to complex ecologies of care in which we circulate. Furthermore, they share a teleology-of-care rationale that assumes that (*1*) there is a health ideal, a consensual understanding of health on which health-care participants agree; and (*2*) there are more or less correct, productive disease-specific paths "toward" health that can and must be followed in order for progress to occur. In this way, compliance expectations, like ableist views of disability, are based on a "utilitarian concept of the body," "the Cartesian image of an individual as a separate, isolated, efficient machine whose goal [is] self-mastery" (Garland-Thomson, 1997, p. 39). The problem of compliance is, however, not one that is most productively or ethically attributed to or located in individuals. Rather, the problem of compliance is that it is an exclusionary fantasy that presupposes an oversimplified account of agency for health-care participants and, in so doing, enacts a teleology of "personal reconstruction through acting on the body" (Rose, 2007, p. 26). And this fantasy is often used to regulate what can count as ethical practices of care: progress down a prescribed path, toward prescribed goals. Anything else is evidence of a failure of mastery or lack of control for both patients and providers, something that requires correction. Because these logics of compliance are historically allied to ableist ideas of agentic action, if we are going to challenge the disabling culture of compliance in biomedicine and its attendant assumptions, we must also challenge the disability system to which they are bound.

Revaluing Noncompliance, Seeing Divergent Paths

In compliance frameworks tied to a disability system that privileges mastery and control, when a person acts in a way that does not follow prescribed pathways deriving from disease-specific diagnostic criteria, the body is—*our* bodies are—often characterized as if they are victims of our moral and psychological shortcomings, lack of cognitive and emotional discipline, and failure to be willful in the right ways. In this way, "willfulness—and other failures of the will," Sara Ahmed (2014) writes, "becomes that which threatens the degeneration of the whole body; not to function would cause the body to become dysfunctional" (p. 111). Indeed, even in the most sympathetic treatment in which the "problem of noncompliance" is defined as a problem of "fit" and the "burden of treatment," noncompliant patients are often characterized as "*failing* to cope" (Mayo Clinic, 2010, emphasis mine). In order to anticipate such potential failures of patients, biomedicine has developed strategies like "predictive modeling" to profile patient personality types as more or less likely to be noncompliant (Chesanow, 2014). For example, a focused literature review of studies that account for the psychology of patient noncompliance in periodontal maintenance concluded that "non-compliance has been linked to negative aggression and immaturity, passivity, dependence and depression, emotion-focused coping, external locus of control and poor coping abilities, low EI [emotional intelligence], neuroticism, and perceived control" (Umaki, Umaki, & Cobb, 2012, p. 399). "Clearly," the authors conclude, "compliance to periodontal treatment is *threatened* by many psychologic variables" (p. 399, emphasis mine).

Sliding from some evidence in measurement studies of a correlation between noncompliance and specific psychological traits to a "threatening" causal relationship, as this study does, is a strategy familiar among such accounts and reflects the easy, common sense assumptions about cause and effect and agency that still too often go unaccounted for. Noncompliance doesn't just correlate with certain perceived personality traits, according to such accounts, but is *caused* by the passivity and other weaknesses of individual patients. As I have noted, such thinking is troublesome for the ways that it is bound to a disability system that has naturalized control and mastery of the body as an ethical imperative. However, such thinking is also harmful for the ways that it contributes to the development of obstacles to treatment one would think even the most pragmatic in the medical community would see as crises, like unused medical advice and ineffective and costly "solutions"—including decades of positivist studies that attempt to solve the "problem of patient communication" by focusing on methods of delivery and strategies for persuading patients to "act on good advice" (Segal, 2005, p. 147). And acting as though compliance were the "norm" against which noncompliance can be defined as deviant and immoral, something to be measured and corrected, has been damaging both to subjects and to systems of care.

As some bioethicists have begun to argue (Murray & Holmes, 2009; Walker, 2009), the presumption of the "autonomous subject" implied by blame and shame rhetoric focused on individuals who do not follow prescribed, agreed-on, or expected treatment and wellness paths has not always served our understanding of ethical patient behavior, nor has it often served, I would add, biomedicine's values of efficiency and accuracy. Contrary to the dominant discussion in medicine about noncompliance, texts (like patient discharge instructions and clinical studies reports) and behaviors (like non/compliance and care of self and others) are not simply the direct consequence of autonomous subjects making choices—perfectly or imperfectly, reactively or proactively. We develop choices, actions, texts, and beliefs—we are rhetorical—as embodied subjects in a material world that is not of our own making. Because our subjectivities are differently circumscribed by our backgrounds, institutional influences, and multiple roles in the world (those that are given, expected, and chosen), as S. Scott Graham (2015) argues, "the practices of health care professionals [and, I would add, participants of all kinds] replicate both structuralist accounts of ideological interpellation and the poststructuralist theories of multivocal subjectivity" (p. 378). Our agency, in other words, is not best understood as either "resistance versus authority or individual versus ideological," and it is not most productively thought of as a *thing* we willfully choose to call forth as autonomous "agents of change" (p. 378), as Graham, following other rhetoric scholars, has argued (Herndl & Licona, 2007; Keränen, 2007; Miller, 2007; Scott, 2006).

"The world kicks back," feminist theorist and theoretical physicist Karen Barad (2007) says to explain her theory of agential realism that explores the material consequences of knowledge. Barad's discussion of the material-discursive ontology of agency and intention has much, I think, to offer to our thinking about the question of rhetorical agency. Specifically, it amplifies why an understanding of agentic action that relies on either a model of agency as socially constructed or one that is premised on the reification of autonomous human agents might be unproductive for developing more just and effective systems of care and, therefore, why our current understanding of non/compliance is dysfunctional. Contrary to traditional humanist accounts of agentic action, Barad posits that agency and other influential phenomena are produced intra-actively—in particular moments, moment *by* moment, over and over again.[13] Arguing for the inseparability of what have historically been distinguished from one another as "the material" and "the discursive" (though the boundary between them hasn't always been obvious or given), Barad addresses the ways that "intention" (her word) "might be better understood as attributable to a complex network of human and nonhuman agents, including historically specific sets of material conditions" (p. 23). Though Barad is clear that she does not mean to suggest that intention, responsibility, and influences are or should be evenly distributed, useful in her account is the insistence on the ontological inseparability of influence and agency and the recognition that agency emerges from material-discursive intra-actions. And she means this literally, as she repeatedly notes that her theory of agential realism is not an "analogical" argument (pp. 21–24, 88, 94): "different material-discursive practices produce different material configurings of the world" (p. 184), Barad writes, and the unique intra-actions in any given moment "not only reconfigure spacetimematter but reconfigure what is possible" (p. 182).

Alongside Annemarie Mol (2008), Stuart Murray (2005), and many posthumanist scholars who have challenged humanist assumptions about agentic subjects, Barad's account of agency as materially and discursively intraactive seems not only relevant to but also important for any ethical project that endeavors to make sense of patient, provider, and health-care system dynamics. It reminds us that we must attend to the ways that the choices and actions of health-care participants (both patients and professionals) are not simply conjured by the will of righteous autonomous subjects but rather are the logical consequences of the dynamic and kairotic convergence of many phenomena.[14] They are embodied, the material-discursive products of "distinct agencies [that] do not precede, but rather emerge through, their intra-action" (Barad, 2007, p. 33). And so, if we are interested in doing the ameliorative work of feminist theory in the contexts of health and medicine, we ought to heed this as a critical reminder that noncompliance—like any phenomenon—can disclose to those paying attention to it something more than just the character of individual patients (their so-called deviance, errant willfulness, and failures as individuals). The divergence of health-care participants can disclose something about systems and contexts for care, care practices, resources, and ways of knowing. Taking this seriously, we ought, therefore, to explore theories that take as their focus something more than reasons to ignore or ways to correct for divergence.

To illustrate the kind of opportunities for care in health and medicine that would be enabled by this kind of engagement and understanding, I offer two brief but compelling examples: (*1*) the support available for people who continue to smoke after being diagnosed with cancer, and (*2*) the therapeutic relationship with patients who are considered to be noncompliant or nonadherent. Though the actual percentages vary depending on the kind of cancer and when the study is conducted (how long after diagnosis), a 2009 study found that nearly 60 percent of smokers diagnosed with cancer continued to smoke in spite of the fact that the prognosis for smokers has long been understood to be worse than for those who never smoked and for those who quit when diagnosed

(Burke, Miller, Saad, & Abraham). However, because the pharmaceutical industry and biomedical community develop drugs and treatment recommendations through a framework that both expects and assumes compliance, two critical opportunities have received less attention. First, the presumption of compliance and a corresponding, selectively minoritizing view of noncompliance have directed attention toward drug development and approval processes that have not taken into consideration the effects of nicotine on the metabolism of anticancer agents and, therefore, have led to missed opportunities to develop treatment that would be more effective for those who continue to smoke after being diagnosed with cancer. As a 2012 review essay reports, "most anticancer drugs do not undergo a formal assessment of their efficacy or toxicity in smokers compared with nonsmokers" and so "little prescribing information is available to guide clinicians on the vast majority of these agents" (Petros, Younis, Ford, & Weed, pp. 929, 920). Though the authors of the essay note that the implications of data from these early studies are still not clear, they conclude that the results of their work "may lead to important questions regarding approaches to smoking cessation in patients with cancer" (p. 920).

While the results of this work may indeed lead to questions related to smoking cessation, such a conclusion illustrates the ways that a culture of compliance has directed attention away from some conclusions and toward others. In other words, this conclusion is, like the study of anticancer agents, limited by persistent, pervasive compliance frameworks and the assumptions on which such frameworks are premised. It is possible that those continuing to smoke after being diagnosed with cancer would benefit from another way of paying attention to evidence of noncompliance in future research other than cessation support: they might benefit, for example, from research that helps develop treatment guidelines that would be more effective for people who, in spite of the fact that cessation might be ideal, will continue to smoke.[15] If future studies pursued this line of research, oncologists who were aware that someone seeking treatment was a smoker could account for the degree of nicotine consumption in the amount prescribed. In other words, the mere acknowledgment of nicotine use among cancer patients, instead of the presumption of compliance, could lead to more effective and efficient decision-making processes and treatment, and given the often tremendous cost of anticancer agents, more appropriate prescriptions could save a great deal of money for both patients and payers.

A related opportunity for crafting more effective and efficient care that is disabled by a culture of compliance—but would be more accessible for knowledge-making inside a framework that recognizes the dynamic constitution of agencies and phenomena—is the opportunity to develop a more open and transparent therapeutic relationship with people who are noncompliant. Even if physicians had the information they would need about anticancer agents to make decisions about treatment that accounted for relevant patient behaviors, a culture of compliance can encourage people to "pass" as compliant, leading some with cancer to underreport their nicotine consumption because they are concerned about losing the support of care providers (Wells, English, Posner, Wagenknecht, & Perez-Stable, 1998). In fact, researchers interested in considering the effects of smoking on people with cancer—and who might, therefore, want to draw conclusions based on data collected about smokers with cancer—are warned that "retrospective use of medical record data on smoking history is discouraged due to inaccuracies in documentation and questionable validity (underestimation) of self-reporting" (Petros et al., 2012, p. 922). And this underreporting of behaviors marked as "unhealthy" or noncompliant is not, of course, unique to people with cancer. International studies have found that self-reported alcohol consumption only accounts for between 40 and 60 percent of alcohol sales (Boniface & Shelton, 2013). This means that, on average, either people are hoarding huge amounts of alcohol or people tend to report that they drink much less than they actually do (about half as much). For this reason, medical school students being taught how to take a patient's history are often advised to estimate that the actual consumption of alcohol can be at least double what a person reports he or she consumes. "The system has trained us well," one medical school student writes. "It's efficient. It's controlled. It's polite. But we are all a little less honest for it" (Yurkiewicz, 2013).

Kairologies of Care

Frameworks for or orientations to care that do not prefigure those who are noncompliant as failing to appropriately care for themselves and others—as immoral or less moral and, therefore, less deserving—might enable more openness and honesty. By recognizing and exploring the "*mutual constitution of entangled agencies*" in self- and other-care (Barad, 2007, p. 33), such frameworks could furthermore explore divergent behaviors not simply as the "fault" of individuals but as a critical yet small part of a much bigger picture. That kind of exploration might be considered a "kairology of care" approach—one that pays close attention to care as situated, embodied, and rhetorical; materially, discursively, and temporally emergent; intra-active; and, therefore, kairotic.[16] Such an approach would, at minimum, most certainly not see the primary purpose of paying attention to divergent behaviors as developing the means and evidence for assigning blame or responsibility for outcomes, or as determining the most persuasive arguments that might be used to modify behavior; rather, it would—following more from the values described by Annemarie Mol (2008) as the "logic of care"[17]—see divergent behaviors as disclosing opportunities for care in specific moments that resist what Stuart Murray (2005) has called the overdetermination of the "rhetorical expression of life" (p. 112). It would consider the contexts for and logistics of specific divergent acts as essential to understanding *that*, *why*, and *how* people diverge from the prescribed or agreed-on behaviors of health-care providers and, therefore, *that*, *why*, and *how* specific intra-actions come to matter for individuals. With regard to the example of people who continue to smoke after being diagnosed with cancer, some of this work might include recognizing that continuing to smoke requires an ongoing adjustment to prescribed medications and an assurance that drug approval processes and protocol are in place that can help support such adjustments. It might also include consideration of why smoking as a coping behavior seems most accessible and desirable to some individuals in some moments and how care professionals, social services, and governments might incorporate that information to offer additional support.

Kairologies of care need to invite and keep track of the reasons why people respond as they do to the challenges of illness, and try to make sense of what convergences of intra-active phenomena can tell us about meeting people where they are when they are there, to design treatment plans, medication schedules, and wellness paths that will work with their needs, goals, and means. They need to recognize that our agency and identity as subjects experiencing the various sensations of health and illness are produced in and by specific sociohistorical and political moments, and that each of those moments are unique articulations, progressive disclosures without distinct beginnings and endings. Such approaches must proceed with an awareness that our agency as health-care participants is not an opportunity for or consequence of fixed decisions we make once and for all (in the past or for the future) and that framing it as such can distract us from what care can be: iterative and flexible, opportunistic, produced and reproduced differently in different moments.

Conclusion

Divergent practices continue to challenge circumscribed notions of rational and scientific, humanist attempts to control outcomes and manage what counts as actionable knowledge in health and medicine. When considered outside a framework that prefigures their ethical value, they can remind us of the ways that care is and should be kairotic, and, therefore, the ways that it is always and already rhetorical. As Thomas Rickert (2013) argues, "kairos is not about mastery but instead concerns attunement to a situation, with attunement understood not as a subjective state of mind or willed comportment but as an ambient catalysis" (p. 98). When people diverge from treatment recommendations or best practices, perhaps the only thing it is safe to assume is that they are opportunistically working with the constantly changing affordances and constraints of their situated, embodied experience. And so a better understanding of how and why people do this in different contexts—and, at minimum, incorporating *that* they do so as "legitimate" (not dismissing as "mere deviance") into guidelines and treatment recommendations—is critical to crafting better systems of care.

If we have learned nothing else from the prevalence and persistence of noncompliance among health-care participants in all areas of medicine by now, it ought to be that paths in health and medicine cannot be fully anticipated or predetermined—and yet our policies and guidelines for care often do not recognize or are not capable of recognizing, accommodating, or working with this. Because resources, knowledge, conditions for care, the experience of illness, and other kairotic details are unstable, instead of assuming noncompliant acts are self-destructive signs of *failures* to cope, we ought to assume that such acts are emergent responses to complex, rhetorical situations in an individual's contexts for care. And, as such, they are actually *evidence of* coping. As health-care participants diverge and converge with prescribed or agreed-on therapies, the variables associated with our specific situations and our experiences of them change. Recognizing this prior to judgment and assigning responsibility is central to improving therapeutic alliances and systems of care. For this reason, care, as Mol (2008) argues, ought to be thought of as an "open-ended process" (p. 22) and "patientism" understood as "exploring ways of shaping a good life" (p. 47).

Instead of treating people who diverge from prescribed or expected practice as "inconvenient" or as obstacles to care, we need to consider that divergent behaviors may be a part of a person's way of "craft[ing] more bearable ways of being" (p. 53) or "shaping a good life," even if how this might be true isn't clear. All behaviors in health and medicine are embodied, contingent, and intra-active and so have something to teach us about ambient influences and the agencies they emerge from and help to produce, and about the convergence and divergence of ideas, practices, affects, bodies, spaces, and moments. To craft more ethical and effective systems of care, we need to invite accounts of and engage with the divergence of all health-care participants (not just patients) and treat them as important feedback mechanisms. We need to pay attention to why behaviors are divergent in the ways that they are and notice the ways that people with different roles, documents, bodies, and technologies intra-actively produce knowledge that might be useful to improving systems of care. Instead of assuming that the ethical response to divergence is correction (in order to persuade convergence), we need to ask, "What kind of ethical action, what kind of care responses might divergence persuade?" At minimum, such an approach can help us amplify the ways that casting the subjects of divergent acts as deviant and destructive undermines the goals of ethical, effective, efficient care. We may not always agree that the actions of people who diverge from best practices are in their best interest or the best interests of their communities, but if we do not recognize the content of people's divergent paths, we forfeit this valuable opportunity for feedback and care.

Notes

1. This is increasingly true, as David Martins (2005) argues, because "once patients were allowed, though not without controversy, to self-administer insulin by injection, they were expected to make daily decisions regarding disease management practices. Compliance-gaining strategies became a critical concern of the medical profession" (p. 60).

2. A 2014 study report in the *New England Journal of Medicine* found that the problem of trust is especially powerful in the United States. In fact, the United States ranks much lower in "trust in physicians" than other countries, according to the article, and class seems to play a role in exacerbating that

> U.S. adults from low-income families (defined as families with incomes in the lowest third in each country, which meant having an annual income of less than $30,000 in the United States) are significantly less trusting of physicians and less satisfied with their own medical care than adults not from low-income families. . . .
>
> Although non–low-income Americans expressed greater trust in physicians than their low-income counterparts did, when responses were analyzed by income group, the United States still ranked 22nd in trust among the 29 countries.

Also interesting in this study is that men in the United States were "significantly more likely than U.S. women" to think that physicians can be trusted (63 percent versus 54 percent) (Blendon, Benson, & Hero).

3. Mel Chen argues in *Animacies* (2012) that the concept of animacy affects how we understand what it means to "care," and I see the concept of compliance in biomedicine as intersectional for similar reasons. Ongoing concerns about compliance are deeply imbricated in and inextricable from other justice-oriented concerns, like those addressed in feminist and queer studies, disability studies, and critical race studies. Chen writes that "categories of sexuality are not color-blind" (p. 104), and the same can be said of characterizations of divergence from agreed-on or expected practices in health and medicine. Compliance expectations are neither color-blind nor are they free of other biases and prejudices. They are—to varying degrees—often ablest, sexist, racist, and classist. While it is beyond the purview of this essay to address all of these critical intersections, I conduct my analysis in this chapter acutely aware that the very premise of "deviance" is one that applies unevenly and that some are more implicated than others.

4. In their introduction to *Material Feminisms* (2008), Stacy Alaimo and Susan Hekman argue that

> feminist theory is at an impasse caused by the contemporary linguistic turn in feminist thought. With the advent of postmodernism and poststructuralism, many feminists have turned their attention to social constructionist models. They have focused on the role of language in the constitution of social reality, demonstrating that discursive practices constitute the social position of women. (p. 1)

While they acknowledge that this turn to the linguistic and discursive has been "enormously productive" and has helped us elucidate the "interconnections between power, knowledge, subjectivity, and language" (p. 1), they note that material reality is often posited by social constructionist theories as a "realm entirely separate from that of language, discourse, and culture" (3). Elizabeth Wilson similarly calls for such work in her 2004 book *Psychosomatic: Feminism and the Neurological Body*, arguing that "the cultural, social, linguistic, literary, and historical analyses that now dominate the scene of feminist theory typically seek to seal themselves off from—or constitute themselves against—the domain of the biological" (8).

5. For important scholarship that addresses bodily rhetorics and the relationship between rhetorics thought to be *of the mind* and those thought to be *of the body*, see *Embodied Rhetorics: Disability in Language and Culture* (Wilson & Lewiecki-Wilson, 2001); *Bodily Arts: Rhetoric and Athletics in Ancient Greece* (Hawhee, 2004); *Rhetoric in the Flesh: Trained Vision, Technical Expertise, and the Gross Anatomy Lab* (Fountain, 2014); and *Disability Rhetoric* (Dolmage, 2014).

6. "Disability history and feminist and disability critiques of science," Dolmage and Lewiecki-Wilson (2010) argue, "provide important methodological lenses, which help the researcher challenge epistemologies that do not self-reflexively examine their own dispositions toward and assumptions about the subject of study" (p. 30).

7. A few clarifying notes about my use of charged or perhaps less familiar terminology: Unless I am referring to the specific phrasing of a particular study or account, I most often use *noncompliance* and *noncompliant* to refer to any behaviors that have been considered to be either "noncompliant," "non-adherent," or "non-concordant" with treatment decisions. I use the phrase *health-care participants* often to refer to people when they are patients or seeking care of any kind as well as to refer to others who are involved in caring for them. I use *non/compliance* to refer to both noncompliance and compliance at the same time and to draw attention to the often fuzzy—yet assumed—boundary between them. And when I use the word *patient*, it is to reference the figure/role of "the patient" as it circulates in biomedicine.

8. In the past ten years or so, Stuart Murray and other bioethicists have challenged the notion of patient autonomy. In their introduction to the 2009 collection *Critical Interventions in the Ethics of Healthcare: Challenging the Principle of Autonomy in Bioethics*, Murray and Dave Holmes assert that

> without an analysis of power and its complexities, mainstream bioethics is ill-equipped to consider social, political, or even economic issues in any robust manner. To begin to wage a critique, we must call into question and resist the reductive binary logic that tends to inform mainstream bioethics: theory/practice, mind/body, subject/object, nature/culture, and so on. (p. 2)

9. To be clear, Martins's relevant argument does not emphasize the problem of locating agency in patients, but it does discuss at length the issue that the increasing prevalence of biotechnologies, the attendant expectations that people adopt more and more self-monitoring and self-care practices, and the common practice of identifying the locus of agency as the purview of individual autonomous subjects have combined to limit research on compliance. Together, Martins argues, they have contributed to an "exclusive focus on interpersonal, dyadic relationships and the presumption that agency can be transferred from one individual to another through an exchange of certified knowledge, skills, and technologies." "Such a model of agency," he writes, "impoverishes what counts as context" (p. 61).

10. Most famously, because it was suspected that some women were abusing the so-called privileges of the sick role to get out of their "domestic duties," the

"rest cure" for women developed by Dr. S. Weir Mitchell was used as a way of testing the genuineness of a woman's claim to illness: "The idea was to provide the patient with a drawn-out experience of invalidism, but without any of the pleasures and perquisites which usually went with that condition" (Ehrenreich & English, 2005, p. 150).

11. This is true in genomics as well: "the body rematerializes," Happe (2013) argues, "when genomics must make fathomable and palpable the body at risk, both to fashion medical subjects and to engage in the making of procedural rules and norms necessary to formalize and routinize particular sets of interventions" (p. 2).

12. One medical anthropologist writes that "the fracture of a . . . femur has, within the world of nature, no more significance than the snapping of an autumn leaf from its twig" (Sedgewick, 1973, p. 31).

13. In Barad's account, a phenomenon is the "primary ontological unit" (p. 141) produced via a *"specific intra-action of an 'object' and the 'measuring agencies'"* (p. 128). Along these lines, noncompliance might be considered a phenomenon that is produced in part in the intra-action of a person responding to illness and the frameworks used for making sense of that behavior.

14. The different inflections of the term *kairos* are many and its rich history is somewhat complex (see Sipiora and Baumlin [2002], for example, for dedicated accounts of this). My use of the word *kairos* here is meant to recall Thomas Rickert's 2013 proposal that "kairos can no longer be thought of as irrational" (p. 97) and Debra Hawhee's (2004) "syncretic" discussion of kairos that draws on the many different inflectional varieties of the word to argue for kairos's capacity to mark the "quality of time" (p. 66). Furthermore, Hawhee's account, which merges the "accommodation" and "creation" models of kairos, also works well for my discussion because it draws attention to the mutual constitution of agency in rhetorical practices.

15. I should note that smoking cessation support is also something that has been identified as a need not often enough accompanying treatment for those with cancer, and it is not my intention to undermine that kind of support. Elizabeth Donaldson (2011) contends that some arguments for recovering disruptive acts and bodies as a form of empowerment can run the risk of minimizing suffering and drawing needed attention away from addressing certain conditions as troubling for their subjects. An argument for noncompliance as a feedback mechanism might similarly be feared to have the potential to draw resources away from supporting those who struggle to follow complex treatment regimens that may be critical to alleviating their suffering. And it is certainly not my goal to undermine anyone's support; on the contrary, it is my

hope that the call for changing the culture of biomedicine to draw forth and recognize accounts of noncompliance without judgment would provide a more open, heterogeneous, and diverse exchange among health-care participants, which would help those who work so hard to educate and support participants managing complex treatment burdens.

16. The term *kairology* was something I first encountered in Judy Segal (2005) and in medical rhetoric scholarship; it has since been picked up by Molly Margaret Kessler (2016) as a way of approaching "how, when, and why certain arguments are persuasive" (p. 238). Though my invocation of the term shares the obvious origin of the concept of kairos and its interest in "contingency" (Segal, 2005, p. 22) and also, to some extent, an interest in "a study of historical moments as rhetorical opportunities" (p. 23), my use of it here is also and primarily meant to provide a way of considering interventions in the current moment, of emphasizing the need to think more about concurrency and the progressive, ongoing emergence of moments and opportunities for care—all of which will become past moments and some of which have yet to happen.

17. The values Mol proposes in *The Logic of Care: Health and the Problem of Patient Choice* (2008) are offered as an alternative to what she calls the "logic of choice"—the logic on which current decisions and care are founded and which emphasizes the ethics of gathering the "best" information in order to make the "right" choices. Logic of care values and practices, according to Mol, include "attempts to craft more bearable ways of being" (p. 53), not taking anything as entirely fixed (p. 61), and seeing that "balance is about attunement" (p. 62).

References

Agency for Healthcare Research and Quality. (2012). Medication adherence interventions: Comparative effectiveness [AHRQ Pub. No. 12-E010-1]. Accessed 1 August 2015. Retrieved from http://effectivehealthcare.ahrq.gov/ehc/products/296/1249/EvidenceReport208_CQGMedAdherence_Executive Summary_20120904.pdf

Ahmed, S. (2014). *Willful subjects*. Durham, NC: Duke University Press.

Alaimo, S., & Hekman, S. (2008). Introduction: Emerging models of materiality in feminist theory. In S. Alaimo & S. Hekman (Eds.), *Material feminisms* (pp. 1–22). Bloomington: Indiana University Press.

Bacall, A. (2011). Cartoon. Accessed 2 July 2013. Retrieved from https://www.cartoonstock.com/cartoonview.asp?catref=aban1045

Barad, K. (2007). *Meeting the universe halfway: Quantum physics and the entanglement of matter and meaning*. Durham, NC: Duke University Press.

Behforouz, H. L., Drain, P. K., & Rhatigan, J. J. (2014). Rethinking the social history. *New England Journal of Medicine, 371*, 1277–1279.

Blendon, R. J., Benson, J. M., & Hero, J. O. (2014). Public trust in physicians—U.S. medicine in international perspective. *New England Journal of Medicine, 371*(17), 1570–1572.

Boniface, S., & Shelton, N. (2013). How is alcohol consumption affected if we account for under-reporting? A hypothetical scenario. *European Journal of Public Health, 23*(6), 1076–1081.

Burke, L., Miller, L.-A., Saad, A., & Abraham, J. (2009). Smoking behaviors among cancer survivors: An observational clinical study. *Journal of Oncology Practice, 5*(1), 6–9.

Chen, M. Y. (2012). *Animacies: Biopolitics, racial mattering, and queer affect.* Durham, NC: Duke University Press.

Chesanow, N. (2014). Why are so many patients noncompliant? *Medscape.* Accessed 30 April 2015. Retrieved from http://www.medscape.com/viewarticle/818850

Conrad, P., & Schneider, J. W. (1992). *Deviance and medicalization: From badness to sickness.* Philadelphia, PA: Temple University Press.

Davis, L. J. (2002). *Bending over backwards: Disability, dismodernism, and other difficult positions.* New York: New York University Press.

Davis, L. J. (2013). *The end of normal: Identity in a biocultural era.* Ann Arbor: University of Michigan Press.

Diedrich, L. (2007). *Treatments: Language, politics, and the culture of illness.* Minneapolis: University of Minnesota Press.

Dolmage, J., & Lewiecki-Wilson, C. (2010). Refiguring rhetorica: Linking feminist rhetoric and disability studies. In E. E. Schell & K. J. Rawson (Eds.), *Rhetorica in motion: Feminist rhetorical methods and methodologies* (pp. 23–38). Pittsburgh, PA: University of Pittsburgh Press.

Dolmage, J. T. (2014). *Disability rhetoric.* Syracuse, NY: Syracuse University Press.

Donaldson, E. J. (2011). Revisiting the corpus of the madwoman: Further notes toward a feminist disability studies theory of mental illness. In K. Q. Hall (Ed.), *Feminist disability studies* (pp. 91–114). Bloomington: Indiana University Press.

Ehrenreich, B., & English, D. (2005). *For her own good: Two centuries of the experts' advice to women* (Rev. ed.). New York, NY: Anchor Books.

Foucault, M. (1973). *The birth of the clinic: An archaeology of medical perception.* (A. M. Sheridan Smith, Trans.). New York, NY: Vintage.

Fountain, T. K. (2014). *Rhetoric in the flesh: Trained vision, technical expertise, and the gross anatomy lab.* New York, NY: Routledge.

Frank, A. W. (1991). *At the will of the body: Reflections on illness.* Boston, MA: Houghton Mifflin.

Fung, B. (2012, September 11). The $289 billion cost of medication noncompliance, and what to do about it. *Atlantic.* Accessed 23 August 2015. Retrieved from: http://www.theatlantic.com/health/archive/2012/09/the-289-billion-cost-of-medication-noncompliance-and-what-to-do-about-it/262222/

Garland-Thomson, R. (1997). *Extraordinary bodies: Figuring physical disability in American culture and literature.* New York, NY: Columbia University Press.

Garland-Thomson, R. (2011). Integrating disability, transforming feminist theory. In K. Q. Hall (Ed.), *Feminist disability studies* (pp. 13–47). Bloomington: Indiana University Press.

Gottlieb, H. (2000). Medication nonadherence: Finding solutions to a costly medical problem. *Drug Benefit Trends, 12*(6), 57–62.

Graham, S. S. (2015). *The politics of pain medicine: A rhetorical-ontological inquiry.* Chicago, IL: University of Chicago Press.

Hall, J. (2006). Facilitating improved compliance among patients with diabetes. *Podiatry Today, 19*(6). Accessed 10 June 2011. Retrieved from http://www.podiatrytoday.com/article/5555#sthash.4V1ew081.dpuf

Hall, K. Q. (2011). Reimagining disability and gender through feminist disability studies: An introduction. In K. Q. Hall (Ed.), *Feminist disability studies* (pp. 1–10). Bloomington: Indiana University Press.

Happe, K. E. (2013). *The material gene: Gender, race, and heredity after the Human Genome Project.* New York: New York University Press.

Hawhee, D. (2004). *Bodily arts: Rhetoric and athletics in ancient Greece.* Austin: University of Texas Press.

Herndl, C. G., & Licona, A. C. (2007). Shifting agency: Agency, *kairos*, and the possibilities of social action. In M. Zachry & C. Thralls (Eds.), *Communicative practices in workplaces and the professions: Cultural perspectives on the regulation of discourse and organizations* (pp. 133–153). Amityville, NY: Baywood.

Keränen, L. (2007). "'Cause someday we all die": Rhetoric, agency, and the case of the "patient" preferences worksheet. *Quarterly Journal of Speech, 93*(2), 179–210.

Kessler, M. M. (2016). Wearing an ostomy pouch and becoming an ostomate: A kairological approach to wearability. *Rhetoric Society Quarterly, 46*(3), 236–250.

Lock, M., & Nguyen, V-K. (2010). *An anthropology of biomedicine*. Oxford, UK: Wiley-Blackwell.

Lubkin, I. M., & Larsen, P. D. (2006). *Chronic illness: Impact and intervention* (8th ed.). Ontario, Canada: Jones and Bartlett.

Martins, D. S. (2005). Compliance rhetoric and the impoverishment of context. *Communication Theory, 15*(1), 59–77.

Mayo Clinic. (2010, September 9). Noncompliance—Victor Montori, M.D. [Online video]. *YouTube*. Accessed 19 April 2014. Retrieved from https://www.youtube.com/watch?v=flcRKdoaiVk

Miller, C. R. (2007). What can automation tell us about agency? *Rhetoric Society Quarterly, 37*(2), 137–157.

Mol, A. (2008). *The logic of care: Health and the problem of patient choice*. New York, NY: Routledge.

Murray, S. J. (2005). Post-textual ethics: Foucault's rhetorical will. *Critical Studies, 26*(1), 101–115.

Murray, S. J., & Holmes, D. (Eds.). (2009). *Critical interventions in the ethics of healthcare: Challenging the principle of autonomy in bioethics*. Aldershot, UK: Ashgate.

Murray, S. J., & Steinberg D. L. (2015). Autopoiesis | Ethopoiesis: Bioconvergent media in the age of neoliberal biopolitics. *MediaTropes, 5*(1), 125–139. Accessed 15 October 2016. Retrieved from http://www.mediatropes.com/index.php/Mediatropes/article/view/22660

Ofri, D. (2012, November 15). When the patient is "noncompliant." *New York Times*. Accessed 3 August 2014. Retrieved from https://well.blogs.nytimes.com/2012/11/15/when-the-patient-is-noncompliant/?mcubz=1&_r=0

Petros, W. P., Younis, I. R., Ford, J. N., & Weed, S. A. (2012). Effects of tobacco smoking and nicotine on cancer treatment. *Pharmacotherapy, 32*(10), 920–931.

Rickert, T. (2013). *Ambient rhetoric: The attunements of rhetorical being*. Pittsburgh, PA: University of Pittsburgh Press.

Rose, N. (2007). *The politics of life itself: Biomedicine, power, and subjectivity in the twenty-first century*. Princeton, NJ: Princeton University Press.

Schwartz, S. K. (2011, February 25). Patient compliance techniques that work. *Physicians Practice*. Accessed 12 May 2012. Retrieved from http://www.physicianspractice.com/patient-compliance-techniques-work

Scott, J. B. (2006). Kairos as indeterminate risk management: The pharmaceutical industry's response to bioterrorism. *Quarterly Journal of Speech, 92*(2), 115–143.

Sedgewick, P. (1973). Illness: Mental and otherwise. *Hastings Center Studies, 1*(3), 19–40.

Segal, J. Z. (2005). *Health and the rhetoric of medicine.* Carbondale: Southern Illinois University Press.

Showalter, J. W., Rafferty, C. W., Swallow, N. A., Dasilva, K. O., & Chuang, C. H. (2011). Effect of standardized electronic discharge instructions on post-discharge hospital utilization. *Journal of General Internal Medicine, 26*(7), 718–723.

Siebers, T. (2008). *Disability theory.* Ann Arbor: University of Michigan Press.

Sipiora, P., & Baumlin, J. S. (Eds.). (2002). *Rhetoric and "kairos": Essays in history, theory, and praxis.* Albany: State University of New York Press.

Topol, E. (2012). *The creative destruction of medicine: How the digital revolution will create better health care.* New York, NY: Basic Books.

Umaki, T. M., Umaki, M. R., & Cobb, C. M. (2012). The psychology of patient compliance: A focused review of the literature. *Journal of Periodontology, 83*(4), 395–400.

Walker, K. (2009). My life? My choice? Ethics, autonomy, and evidence-based practice in contemporary clinical care. In S. J. Murray & D. Holmes (Eds.), *Critical interventions in the ethics of healthcare: Challenging the principle of autonomy in bioethics* (pp. 15–32). Aldershot, UK: Ashgate.

Wells, A. J., English, P. B., Posner, S. F., Wagenknecht, L. E., & Perez-Stable, E. J. (1998). Misclassification rates for current smokers misclassified as non-smokers. *American Journal of Public Health, 88*(10), 1503–1509.

What doctors wish their patients knew: Surprising results from our survey of 660 primary-care physicians. (2011, February). *Consumer Reports.* Accessed 22 September 2014. Retrieved from http://www.consumerreports.org/cro/2012/04/what-doctors-wish-their-patients-knew/index.htm

Wilson, E. A. (2004). *Psychosomatic: Feminism and the neurological body.* Durham, NC: Duke University Press.

Wilson, J. C., & Lewiecki-Wilson, C. (Eds.). (2001). *Embodied rhetorics: Disability in language and culture.* Carbondale: Southern Illinois University Press.

Yurkiewicz, S. (2013, March 13). "Good patients" cover their emotional cracks. *Scientific American.* Accessed 22 September 2014. Retrieved from http://blogs.scientificamerican.com/this-may-hurt-a-bit/2013/03/13/good-patients-cover-their-emotional-cracks/

Zimmerman, R. (2012, August 31). Patient angst: When you just have to say "no" to the doctor [Post by Dr. Annie Brewster, guest contributor]. *WBUR's Common Health.* Accessed 15 September 2015. Retrieved from http://commonhealth.wbur.org/2012/08/patient-angst-when-you-just-have-to-say-no-to-the-doctor

4.
Mattering Gender

Technical Communication and Human Materiality

Jennifer Bay

By this point in the collection, readers may already be familiar with terms like "agential realism," "new materialism," "object-oriented ontology," and many other concepts derived from the current interest in what might be termed posthuman rhetorics. I want to shift theoretical gears and come at feminist rhetorical science studies from a different angle, asking: what is at stake in these arguments for how we communicate technical or scientific information, and, specifically, how do those stakes affect the budding professionals we teach in our courses? Whether we teach required "service" courses or classes for majors in professional and technical writing, our methods, theories, and approaches show students what is possible in the world and how they can position themselves with respect to those possibilities. Thus, the need for feminist rhetorical science studies derives from the need to pay not less but *more* attention to the human: looking at the nonhuman elements of rhetorical interactions allows us to understand what it means to be human *differently*. Disrupting the classic subject-object binary discloses new worlds and new possibilities for us, especially at a time in which "unprecedented things are currently being done with and to matter, nature, life, production, and reproduction" (Coole & Frost, 2010, p. 4). How, then, can feminist new materialist theory impact the students we teach, many of whom will enter tech industries that are stereotypically masculinist and encounter new technological and scientific practices, procedures, spaces, and approaches that challenge what it means to be human?

Karen Barad's work provides us with an entry into these issues and can be used as a framework for reconsidering how we see the human in new materialist networks. While many scholars have engaged with new materialism, both broadly (Bennett, 2009; Coole & Frost, 2010; Dolphijn & van der Tuin, 2012) and specifically in rhetoric and writing (Barnett & Boyle, 2016; Gries, 2015; Rickert, 2013), few have taken a specifically interdisciplinary, STEM-based feminist perspective like Barad. Barad (2012), for instance, has advanced her work on agential realism to explore the inhuman, introducing the inhuman as another fold in the worldly phenomena we examine and the apparatuses by which we understand them. She argues that

> it is the inhuman . . . that may be the very condition of possibility of feeling the suffering of the other, of literally being in touch with the other, of feeling the exchange of e-motion in the binding obligations of entanglements. (p. 219)

And she further explains in a footnote:

> The inhuman is not the same as the nonhuman. While the "nonhuman" is differentially (co-) constituted (together with the "human") through particular cuts, I think of the inhuman as an infinite intimacy that touches the very nature of touch, that which holds open the space of the liveliness of indeterminacies that bleed through the cuts and inhabit the between of particular entanglements. (p. 222, f19)

In other words, the inhuman does not fit within subject or object but holds open the space in between, living within the intersections of our human-nonhuman entanglements. In some ways, it is the fluidity between the two that enables communication and connection, but it also allows for inequities, hierarchies, and different techniques of power. Where does this inhuman show up in our methods for examining phenomena in the field? What is the impact of our methods on the subjects we deal with on a daily basis? While some theorists may claim to have transcended the subject-object dichotomy in their work, there remain live bodies who are affected by our epistemological and ontological leanings in the classroom. We must still attend to practice.

Technical Communication Theory has often acknowledged nonhuman materialities as inherent to workplace and technical situations, which often involve scientific practices, technologies, and structures. Much of our writing concerns technological objects; technical and scientific communication, for instance, involves writing about scientific objects or new technologies and communicating knowledge about them to human audiences. More recently, theorists have acknowledged that relations between humans and these objects are more complex than at first blush. Actor-Network Theory (ANT) and Activity Theory (AT) have been especially useful for understanding the interconnected relations among technology, rhetoric, and human users, and scholars such as Clay Spinuzzi, David Russell, and others have applied iterations of these theories to varied workplace writing situations. Under ANT, for instance, both humans and objects function as actors in a larger network. The rhetorical situations involving technological objects illuminate the agential roles those objects have in constructing meaning and advancing knowledge. What's important here is that the lines between human and nonhuman are not so clear; drives and forces are blurred and enmeshed with one another to produce action and meaning. As Bruno Latour (1999) writes, "humans are no longer *by themselves*" (p. 190).

Dorothy Smith (1974) and feminist standpoint theorists long ago reiterated the point that the researcher plays an important role in the inquiry process. Smith's 1974 seminal work calls on the discipline of sociology to acknowledge gendered experiences in the position of the researcher:

> To begin from direct experience and to return to it as a constraint or "test" of the adequacy of a systematic knowledge is to begin from where we are located bodily. The actualities of our everyday world are already socially organized. Settings, equipment, "environment," schedules, occasions, etc., as well as the enterprises and routines of actors are socially produced and concretely and symbolically organized prior to our practice. (p. 11)

Smith's work acknowledges that as researchers we each observe from a specific standpoint, and before her work, sociology—like the natural sciences—assumed objectivity on the part of the researcher. However, while sociology and other social sciences that employ qualitative methods have embraced the role of the researcher in the making of knowledge, the "hard" and natural sciences have not fully explored this idea, nor have they examined the role of the instrument in the creation of phenomena. As Latour and Steve Woolgar (1979) and Ian Hacking (1988) point out, instruments allow phenomena to show up for us:

> It is not simply that phenomena *depend on* certain material instrumentation; rather, the phenomena *are thoroughly constituted by* the material setting of the laboratory. The artificial reality, which participants describe in terms of an objective entity, has in fact been constructed. (Latour & Woolgar, 1979, p. 64)

Barad and other new materialist theorists extend this observation to claim that the material setting of the laboratory involves other materialities beyond formal scientific instruments. This includes bodies—or the lack thereof—social instruments, and other factors seemingly outside of the technical apparatus.

While these approaches have been useful for understanding the interconnectedness of technologies, bodies, and discourses, we also struggle with the fact that not all humans have been given equal status in the world, particularly in our academic disciplines. Pulling from Karl Marx, cultural studies first brought this fact to the foreground in the 1980s, and there have been some scholars within Technical Communication who have taken up this call. For instance, Jennifer Slack, David Miller, and Jeffrey Doak (1993) have provided us with a framework for understanding the dynamics of power and culture involved in authorship in the field. Angela M. Haas (2012) has turned our attention to decolonial theories and methodologies. But overall, there has been scant attention paid to the inequities of race, class, and gender. So, while cultural studies has opened up possibilities for looking at how cultural factors impact the production, reception, and circulation of professional and technical writing, "only a few scholars within our field are doing work specifically and explicitly at the intersections of race, ethnicity, rhetoric, and technology studies" (Haas, 2012, p. 281). One problem that might need to be tackled is how we view the subject in the field. Although Slack et al. address authorship and other writers such as Steven Katz (1992) examine how writers present themselves as authors, the status of the *human* subject has not been interrogated. Indeed, many of our theories of the subject assume that features of the human are stable and not culturally specific or disempowered. Latour's human, for instance, is often *assumed* to be fully and generically/genetically human, with gendered, racial, and cultural materialities having no effect on the subject's ability to be considered fully human. Humanness is equal and interchangeable. Furthermore, some of his oft-cited examples orient us toward one specific subject. The speed bump, for instance, only takes into account a human in a car. It doesn't take into account a pedestrian or a school crossing guard, nor does it account for the material specificities of the driver or other nonhumans. These limitations negatively impact the potential for our methods and metaphors to better account for the material specificities of rhetorical interactions. How, then, do we embrace this new turn toward the nonhuman *when the human being has not been given full attention in all of its specificities*?

This chapter charts the potential impact of how studying the nonhuman can illuminate and complicate theories of the subject in Technical Communication. I argue that feminist new materialism can bridge the impasse between the human and nonhuman via the attention it pays to the inhuman, or the irrational and immeasurable aspects of Technical Communication that we know or feel but cannot always materialize. As Susan Hekman (2010) explains, "identities disclose a world for us" (p. 106), but that world is not always visible because of the ways in which we understand subjectivity, situation, and method. In *The Material of Knowledge*, she argues that feminist theory can help solve the stalemate between the linguistic and the material via the shift from epistemology to ontology. Instead of the postmodernist focus on how discourse constructs our world, feminist theories, such as those articulated by Barad and Nancy Tuana (2001, 2008), are illustrating how that discourse gives matter agency: "The agency of matter structures the reality of our world. It sets the boundaries of that reality, establishing what is included and excluded from the real" (Hekman, 2010, p. 74).

In what follows, I trace a few different strands that might demonstrate what is at stake in looking at the field and our research through the lens of feminist new materialism. In the first case, I discuss practical research I completed with an undergraduate student researcher on female practitioners in technical communication in order to explore the material realities of the female students in my classes. This research, however, failed to account for those gendered realities, and I speculate that one of the reasons was that my methods could not account for the material complexities of gender. In the second case, I explore what research in the field might look like using feminist new materialism by comparing two studies, one by Clay Spinuzzi on coworking that uses AT, and another by Susan Nordstrom on qualitative interviewing, which uses what Barad (2003) calls agential realism. I argue that in order for certain complex materialities of gendered existence to show up in our research, we need to adopt and adapt feminist new materialism into our research practices.

Gender Matters in the Undergraduate Curriculum

Since 2003, I have taught in the undergraduate professional writing major program at Purdue University, where I also consistently teach a capstone internship course. Throughout this time, generally more women than men undergraduate majors have been enrolled in the program (we do not have specific data to support this, but I can say with general certitude that at least two-thirds of the undergraduates I have taught are female). Perhaps this is because the professional writing major is housed within the English department, and we know that English majors are stereotypically female. Or it may be that our undergraduates are much more practically focused and have been advised to enter a more practical degree program. Our professional writing degree is oriented around course and client projects, as opposed to our communications degree, which is known to students as being more theoretical and lecture-based. At the university, we also have students who start out in STEM fields and later either transfer to the liberal arts or double major in them. At any rate, I consistently encounter undergraduate women who lack the confidence to see themselves as experts, especially with regard to technology. These women often see their future careers as book or magazine editors, which to them involves reading and evaluating print-based manuscripts. More recently, I have noticed that these students seem more confident with their social media use, but most social media sites are not generally classified as STEM technologies.[1] Sometimes, they aren't quite sure what they will do with their degrees. I distinctly recall one bright, tech-savvy female student who I encouraged to respond to a second interview request with a high-tech company. She politely declined and instead chose to stay locally at her low-paying retail job. When I look at our alumni on LinkedIn, I find many of the women with degrees in professional writing are engaging in other kinds of work, often in what are perceived to be nontechnical fields. I often wonder why these students do not pursue the work for which they have been trained.

I start with this background because as a scholar and teacher, I am constantly on the lookout for research that addresses how we can better prepare and mentor our female students. I seek to understand the lived experiences of the female students I have taught and mentored and how those experiences dovetail with the theory that we study. What are their experiences with the field, with the technologies they use, and with the theories they learn? What happens to them after they leave the program and develop as professionals? We have few textbooks or clear studies about how professional and technical writers, in all their multiplicity, progress through their careers, and we certainly have none about how women specifically navigate the gendered complexities they might face in the workplace and in the world.

While there were many academic articles on gender and technical communication published in the 1990s, until fairly recently, little attention has been paid to the gendered experiences of practitioners on the job. Most of the academic articles that have been published on gender have not addressed the lived realities of female practitioners, nor have they addressed specifically feminist teaching practices that might present ways of understanding the complex racial and gendered dynamics of workplace writing (and how to teach students to navigate those dynamics). With my then-undergraduate researcher, Trinity Overmyer, we took a twofold approach to see if we could uncover some of the gendered complexities that our undergraduate students might face as they develop their identities as professional and technical writers.

We wanted to use previous methods that had been used in the field in order to establish credibility and continue the same scholarly thread. We performed a keyword analysis of how academic and practitioner journals discuss gender in technical and professional writing journals, thinking that there might be connections between gender issues in the classroom and gender issues among practitioners. We also solicited data from a contact at the U.S. Census Bureau to see whether the predominantly female population in our major was reflected in larger employment trends among technical communication practitioners, especially whether women technical communicators were present in heavily male-dominated STEM fields.

We started by adapting Isabelle Thompson's (1999) keyword search on gender as modeled in her article "Women and Feminism in Technical Communication: A Qualitative Content Analysis of Journal Articles Published in 1989 through 1997." Thompson searched five journals in the field and found forty articles on women and feminism over a period of nine years, and most of these articles appeared in special issues on women. "Furthermore," she points out,

> the common themes in these articles were predictably about inclusion of women in technical communication—through eliminating sexist language, providing equal opportunity in the workplace, valuing gender differences, recovering historical contributions by and about women, and critiquing commonly accepted concepts and terms. (p. 155)

Thompson concludes that "it is difficult to ascertain the effects of this research on workplace or classroom practice" (p. 175).

We replicated Thompson's keyword search to see if there was a significant increase in feminist scholarship in the journals after her article appeared *or* if there were more articles providing specific connections to workplace or classroom practices. What we found was that her work did not necessarily prompt an increase in either kind of scholarship. Since 1997, for instance, the *Journal of Business and Technical Communication* has published two articles with *gender, women,* or *feminism/ist* in the titles: "Gender and Modes of Collaboration in an Engineering Classroom: A Profile of Two Women on Student Teams" by Sandra Ingram and Anne Parker in 2002 and Jeanne Weiland Herrick's "'And Then She Said': Office Stories and What They Tell Us about Gender in the Workplace" in 1999. Compare this to a keyword search of these terms in *Management Communication Quarterly*, a journal more oriented on practical workplace issues and not included in Thompson's study. *Management Communication Quarterly* published fifteen articles with *gender, women*, or *feminism/ist* in the title (although this is not to say that the journal is especially attuned to gender, as from 1987 to1997 there were only seventeen articles with *gender, women*, or *feminism/ist* in the title). Since Thompson's five journals focused more on scholarly research and women's academic issues than on practitioner experience, we wondered whether going to specific data about women in the workplace would show something different.

In the second part of our research, our contact at the U.S. Census Bureau provided us with data sets showing trends in employment among technical writers, including gender and status. We juxtaposed these data sets with the data from journals to see whether there were employment trends that dovetailed with the lack of attention paid to gender issues or whether the issues were just not visible. As figure 4.1 shows, almost 60 percent of all technical writers are female while approximately 40 percent are male. However, there are four times the number of part-time female technical writers than male, albeit that constitutes only 8 percent. Interestingly, there are twice as many female tech writers with only a high school degree and significantly more female tech writers with more than five years of college, as can be seen in figure 4.2.

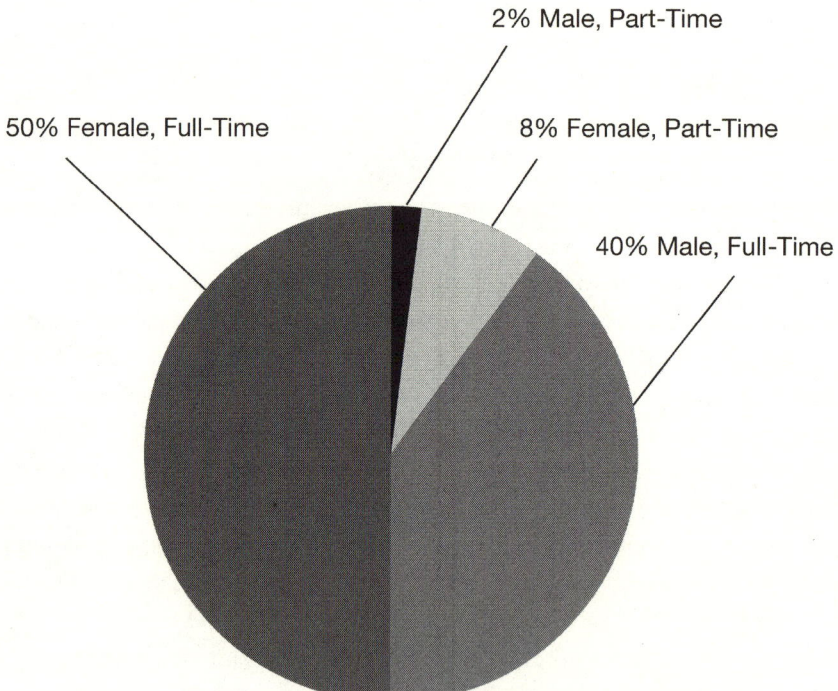

Figure 4.1. Technical writers employed in 2010, by gender and status (Ruggles et al., 2010)

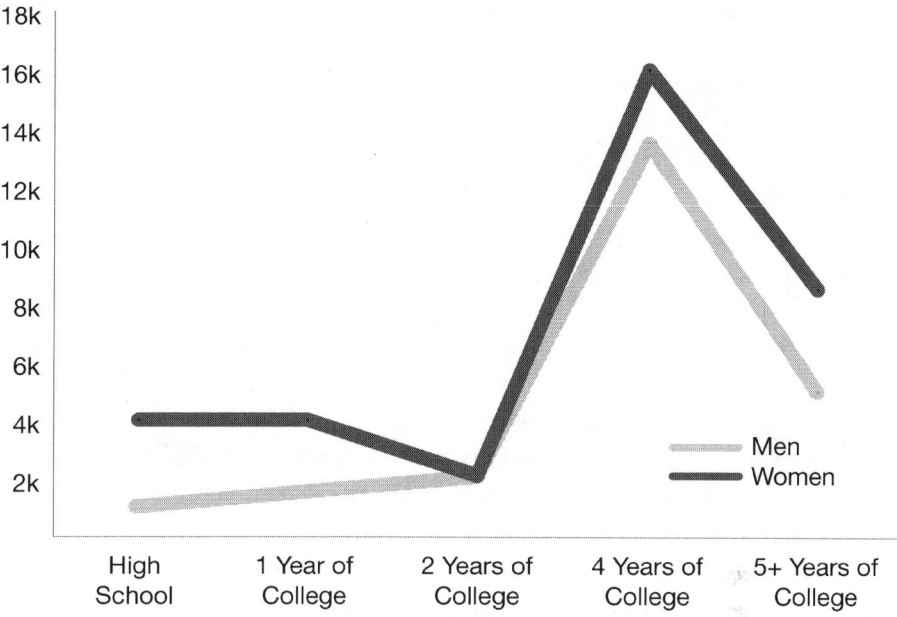

Figure 4.2. Gender of technical writers in 2010, by education levels (Ruggles et al., 2010)

We were surprised at these data, as we assumed from our own experiences with both the large numbers of women in our courses and their technological reticence that there would be fewer female technical communicators in the workforce. How was it that our female students seemed to show such reserve toward the technologies they would use in what we would call technical and professional writing, yet there were more female tech writers in the workforce? If there are more female practitioners, why is there not more attention paid to issues of gender in our scholarship? What kinds of specific experiences do women in the field have on the job, especially in terms of interacting with subject matter experts in tech fields?

We ultimately concluded that our work called for "further research on how female technical writers who work in male-dominated industries are able to successfully navigate the complex gender dynamic" (Overmyer, 2013, p. 98). Our research was completed in 2013 before the more recent focus on gender and racial disparities in Silicon Valley, as well as controversies such as GamerGate, both of which have refocused attention on the issue of gender discrimination in high-tech industries and environments. Even with this new awareness of difference, we still wonder how female tech writers interact with the male-dominated subject matter experts in these fields.

What I realize now is that our research would have benefited from a feminist new materialist approach. We were looking for statistical evidence that there was a gender gap or issue in the field. I had a *feeling* from my teaching that something was happening, but the data did not necessarily show it. I would argue that the phenomenon we were looking for was not apparent in the data because it was invisible to the theoretical lens we were using. We used traditional terms in Technical Communication and traditional understandings of what technical communication is (common search terms), and, as Katherine Durack pointed out in 1997, "notions of *technology, work*, and *workplace* may be gendered terms" (p. 249). While Durack focuses on the history of Technical Communication, she also notes that a stable concept of workplace is disappearing (p. 257). That is, our understanding of "workplace" and "technology" as objects is shifting along with the contexts for writing and communication. Moreover, our methods do not always allow for our relationships with material artifacts to show up. A feminist new materialist approach would acknowledge how those material artifacts affect and are affected by users and technical communicators, forming a new understanding of what technical communication is. Because technology is not a stable object, it shifts and moves, and accordingly, our interactions with it change us.

Thus, articles such as "Redefining the Workplace: The Professionalization of Motherhood through Blogging" by Emily January Peterson (2014) force us to rethink traditional terms in the field (such as *workplaces* and *technologies*) in order to acknowledge the contributions of female technical communicators. Peterson "explores how mom bloggers claim a professional space in communication, and, in so doing, [how] they offer implications for improving professional values and challenge the field to include them as professionals" (p. 278). We've seen this in other recent work, especially in historical studies on the previously unknown contributions of female technical communicators. Edward Malone (2013, 2015), for instance, explores how women such as Elsie Ray and Eleanor McElwee founded our professional organizations via their organizational efforts. Sarah Hallenbeck (2012) discusses technical manuals written by women about bicycling in the nineteenth century. All these works share the idea that extra-organizational—"extra-institutional," as Miles Kimball (2006) terms it—work should be part of what we call professional and technical writing. As Kimball (2006) contends, "we should continue efforts to broaden teaching so students can learn not only to participate in strategic workplace communication, but also to seize control of the possibilities for tactical technical communication, both within and outside of the workplace" (p. 83). Such an extra-institutional approach will not only help us understand our own field "but also the relationship between technology, discourse, and people's lives" (p. 84).

In another sense, we might look at extra-institutional or extra-organizational activities as forms of distributed work, which Clay Spinuzzi (2007) terms "coordinative, polycontextual, crossdisciplinary work that splices together divergent work activities (separated by time, space, organizations, and objectives) and that enables the transformations of information and texts that characterize such work" (p. 266). Spinuzzi identifies this trend as part of the new work economy in which corporate downsizing, entrepreneurship, and mobile practices have connected to form different kinds of networked interactions on the job. I would argue that women, the working poor, and other disenfranchised groups have long been engaged in distributed work practices, especially the kind of extra-organizational activities that do not merit attention. While everyone multitasks, there is a distinctly gendered component to it in the lives of women. I would guess that many of the women reading this chapter right now are also folding laundry, responding to a student's e-mail, writing a thank-you note, caring for an elderly parent, working at a soup kitchen, or otherwise engaging in the kinds of extra-organizational care work for which we feel responsible.[2] The material-discursive reality of many women's lives is that work has been and will always be distributed across a network of embodied concerns and relations.

So, how do our students negotiate what will be (and for many, already are) their embodied realities when they graduate? In our courses, do we convey the *material a/effects* of work to our students, or do we rely on clearly objectivist and scientific understandings of professionalism or what counts as part of the field? Do we discuss the ways in which bodies and technological objects might interact differently according to gender, race, class, and other factors? Perhaps the way we teach our methods in classes impacts the way that students see themselves and their possible career trajectories.

In the next section, I look at two approaches to what could be called "distributed work" in Technical Communication: one is an application of feminist new materialism to qualitative research practices on extra-organizational work, and the other is a traditional study on distributed work processes in technical communication. I juxtapose these two approaches not to privilege one or the other but to demonstrate what can be gained from a feminist new materialist approach to our work. While both are diffusive approaches that work against the subject-object dichotomy, they provide us with a perspective from which to see the human being differently.

Mattering Gender and Objects

Two of the most prominent methodological approaches in Technical Communication are ANT, an approach often associated with Bruno Latour, and AT, developed by Yrjö Engeström and discussed heavily in the field by David Russell (1997). Both are useful for attempting to transcend the subject-object distinction and to illustrate how networked processes and activities emerge via interactions between humans and nonhumans. Both approaches share certain qualities. Brian McNely, Clay Spinuzzi, and Christa Teston (2015) classify both ANT and AT as associative theories, which

> analyze humans and nonhumans as parts of intersubjective systems across which agency and motives are stretched. Such theories do not necessarily deny individual agency or cognition, but they deemphasize the roles of individual human beings to avoid overdetermining human agency and underdetermining roles played by other parts of the system under consideration. (p. 4)

McNely et al. (2015) distinguish associative theories from new materialist ones, which "directly engage the missing masses of nonhumans, taking seriously their potential role in affecting human work, and effecting ostensibly human activities and outcomes" (p. 5). Agency is distributed across humans and things. For instance, Latour argues that we must attend not just to human beings in situations involving technology but also to the technology itself and how it functions as an equally important actant in the network. Thus, both associative and new materialist theories account for the nonhuman elements in any networked process, including nonhuman objects, beings, documents, and technologies; however, associative theories tend to privilege processes and networked relations, whereas new materialist theories tend to spotlight the nonhuman elements in order to see what other relations might be possible. Such an emphasis allows us to also see the human from the perspective of the nonhuman (or what such a perspective might look like). This not only orients our gaze toward the network, but it also allows us to see the human in a different light.

To illustrate the differences between associative and new materialist theories, I focus here on scholarship by Spinuzzi that has advanced the field's understanding of how AT can be taken up in Technical Communication, specifically his use of fourth-generation AT to examine phenomena such as coworking. I focus on Spinuzzi's work for a moment not to critique it but to open it up to other possibilities available through feminist new materialist methods.

Spinuzzi's essay on coworking (2012), part of a book-length study on new work habits, explores new spaces where professionals interact and innovate in Austin, Texas, a city with a tremendous number of start-up technology innovators and incubators. Coworking has received a good bit of attention as a new form of work space for entrepreneurs, small businesses, and independent contractors, emerging from the idea that work is no longer isolated in a physical, often corporate space. In his review of coworking literature, Alessandro Gandini (2015) argues that

> coworking environments provide a space for urban-based freelance, often precarious workers to reterritorialise the physical organisational structure previously offered by firms, which are now diminishing from the emergence of a well-delimited new spatial organisation but with flexible boundaries and affiliations. (p. 198)

The distributed work that is happening in coworking spaces is part of rethinking how work happens, the spaces in which it happens, and thus the kinds of work that can occur.

In "Working Alone Together," Spinuzzi (2012) reports on a two-year study of nine coworking spaces in Austin, interviewing coworkers and coworking proprietors as well as analyzing documents. He argues that "understanding coworking—in particular, how people define it, who decides to engage in it, and why they do it—can help us to develop theoretical and analytical tools for understanding other cases of distributed work" (p. 400). Specifically, Spinuzzi explored these facilities and interviewed the owners, classifying the sites into three kinds: community work spaces, unoffices, and federated work spaces. Community work spaces saw themselves as serving their local communities and offered many different community services, often for free. Unoffices encouraged discussion and interaction, but did not require it. Federated work spaces were characterized by entrepreneurship and formal collaborations. Spinuzzi tells us that his extensive data show disagreement among proprietors and coworkers in each aspect of coworking studied, "the object (what), actors (who), and outcome (why)" (p. 409). A fourth-generation AT approach, he writes, provides us with a way to make sense of the inconsistencies in the data on coworking. "Examining how these people work alone together," Spinuzzi argues, "prepares us to better apply activity theory to other examples of distributed, interorganizational, collaborative knowledge work" (p. 434).

What's of interest to me here is the focus on *overcoming* inconsistencies and contradictions to make meaning and on classifying types of spaces. Feminist new materialism would embrace those inconsistencies and contradictions and argue that it is within them that meaning emerges. Each iteration of coworking that Spinuzzi examines is its own entanglement of humans, nonhumans, spaces, and objects—in short, what could be called a mangle of elements. Susan Hekman (2010) relies on Andrew Pickering's concept of the mangle to theorize complex relations between humans and nonhumans from a feminist perspective. Distinguishing the mangle from ANT, Hekman (2010) notes that the mangle focuses more on the entanglement and the impurity of such an entanglement: "Mangles mix everything up. Diverse elements are thrown into the mangle, and as a result, diverse results emerge" (p. 23). Hekman privileges the mangle as a way to understand the entanglements of humans and the material world for various reasons, but most importantly because "science, its theory and practice, nature, machines, technology, and politics all interact in the mangle. *Mangle* is both a noun and a verb" (p. 24). In short, "what the mangle gives us is an image of how we are located in the world and how the elements of that world interact" (p. 25). The "we" in this quotation is of particular interest, as Hekman goes on to explain that the mangle is specifically useful for examining women and the bodies that have come to define them. Such a move is indicative of feminist new materialism, which focuses on how gendered experience differently constitutes a particular mangle.

But mangles are not independent of the spaces from which they emerge, and space plays an important rhetorical function in the construction of these mangles. As Danielle Endres and Samantha Senda-Cook (2011) argue, place is a rhetorical phenomenon, and as such, the instruments and approaches we use to examine it help to construct it. While the focus of Endres and Senda-Cook's article is on place and social movements, they demonstrate that "the confluence of physical structures, bodies, and symbols in particular locations construct the meaning and consequences of a place" (p. 276). This also includes the kinds of bodies that inhabit (or don't inhabit) those spaces and the procedures enacted within and on a space. "Place subtly changes as bodies move through it," they write,

> and those bodies make choices (often unconscious) about whether to conform to the expectations of place or transgress those expectations. Yet we can and should recognize the embodied and experiential aspects of place even in studies that do not focus on transgressions of the meaning of place. (p. 278)

A feminist new materialist approach, then, would acknowledge how space functions as an apparatus that allows a rhetorical phenomenon to show up.

To pull an example from Spinuzzi, several of the coworking spaces he discusses were located in historical buildings in Austin or in spaces with extra-institutional histories. For instance, the coworking site Brainstorm, Spinuzzi (2012) explains, "was based in a Victorian house in East Austin next to the freeway" (p. 411). Another site, Space12, "was once a notorious East Austin nightclub before a local church took over the site. The church now served the community in which the building is located" (p. 410). In each of the nine sites, coworking *emerged* just as much from the materialities of the sites as from the proprietors and the communities in which they are situated. One of the factors influencing coworkers, for instance, is location, which makes Brainstorm, the site near the freeway, attractive to certain kinds of people. The fact that it is a Victorian house also contributes to its ambiance and the people/workers who might be attracted to it. The space itself has just as much influence on the emergence of the site as the philosophy of the owners or their coworkers. In short, I read each of these sites as a mangle that may have some similarities with other coworking sites, but those buildings and their histories exert just as much agency as their owners and coworkers. In short, the material spaces, like bodies, technologies, and processes, allow coworking to emerge as a rhetorical phenomenon. Coworking, then, as distributed work emerges out of each specific material-discursive interaction, or mangle, and while we might be able to tease out some generalizations, we cannot necessarily isolate and codify practices around them.

Let's look now at what I would call another instance of distributed work, but one that shows up because the researcher takes a feminist new materialist approach, specifically deploying agential realism in her scholarship. Susan Nordstrom's scholarship looks at how material objects impact and affect research practices. Nordstrom (2013) proposes what she calls the "object-interview," "a Deleuzian space in which subjects and objects, living and non-living, entangle together" (p. 237). The object-interview challenges the conventions of the traditional qualitative study, which examines subjects from a humanist perspective. In that model, the subject is a stable, coherent, and independent actor who exerts autonomy in a situation. Nordstrom's work on the family objects of genealogy prompted her to experiment with a different methodology, one that would not "draw information *from* people in order to produce knowledge *about* people and try to understand the meaning they make of their lived experiences" (p. 243). Rather, the object-interview takes both human beings and objects as equally productive agents, producing "an entangled conversational interview of . . . objects, subjects, events, and *a life*"

(p. 243). The object-interview is thus an ontological event, rather than an epistemological one. The object's material being, and not just what we know and say about it, is given a place in the mangle.

Nordstrom (2015) puts her methodology to work in "Not So Innocent Anymore: Making Recording Devices Matter in Qualitative Interviews," where she describes four irruptive moments in her research on the objects of family genealogy and how women genealogists construct family histories from those objects. These moments were irruptive in that the recording devices she was using in her interviews came to matter differently and disrupted "the spell of the objectivist-realist material-discursive practice" of qualitative research (p. 392). In each of these instances, the recording device worked with the body of the researcher, the subject, other materialities, and the physical space to produce the interview itself. During one interview, for example, the subject was noticeably uncomfortable being recorded, so Nordstrom hid the recording device in a stack of papers, in her inner elbow, and in her back pocket as she walked around the house with her subject. She writes,

> The digital recorder picked up the shuffling of papers, the muffling of the microphone partially covered by a thick sweater, and the swish-swash sound of my buttocks in corduroy as I walked around her house. Our conversation undulated on these sounds. I could not deny that the object-interview rode on a sea of attitudes about recording devices and the materiality of the recording device, paper, and my body. (p. 392)

What Nordstrom seems to be describing here is an entangled world of bodies, objects, artifacts, and recording technologies—in short, a mangle that is interconnected via what Barad (2012) might call the inhuman. Nordstrom (2015) writes,

> Each object-interview articulated its own delicately entangled world of humans (both living and nonliving), nonhuman (e.g., family history genealogy objects and recording devices), culture, and discourse. In each world, meaning was no longer a property of people or objects. The entangled world generated meanings that were arbitrarily bound by recording devices that attempted to grasp those meanings. (p. 393)

The devices we use to try to capture the phenomena we study are always already implicated in our research. They become part of the mangle of what we are studying and we cannot "see" without the apparatus, which in this case is the recording device.

Barad's agential realism is key here in informing Nordstrom's work as "an epistemological-ontological-ethical framework that provides an understanding of the role of human *and* nonhuman, material *and* discursive, and natural *and* cultural factors in scientific and other social-material practices" (2007, p. 26). Agential realism, Barad explains, does not examine "independent objects with inherent boundaries and properties but rather *phenomena*" (p. 139). For Barad, "*phenomena are the ontological inseparability/entanglement of intra-acting 'agencies.'*" (p. 139). These intra-actions are the material practices and arrangements that divide or separate subject from object. In simpler terms, Barad tells us that phenomena are constitutive of reality; there are no things in themselves, but things in phenomena. How we understand phenomena is conditioned by the lenses through which those phenomena come into being; in the case of science, the technological apparatus of the microscope allows those phenomena to show up. In the field of Technical Communication, we might understand our methodology as the technological apparatus. As Barad describes,

> apparatuses are not merely about us. And they are not merely assemblages that include nonhumans as well as humans. Rather, apparatuses are specific material reconfigurings of the world that do not merely emerge in time but iteratively reconfigure spacetimematter as part of the ongoing dynamism of becoming. (p. 142)

Thus, for Barad, we also have to consider the apparatus by which we measure and perceive reality. We might see our methodologies, then, as "*dynamic (re)configurings of the world, specific agential practices/intra-actions/performances through which specific exclusionary boundaries are enacted*" (2003, p. 816). That is, it is the approaches we use to examine phenomena that help create the worlds we understand and inhabit.

To return to my "failed" attempt to locate the gendered experiences of women technical communicators, my coresearcher and I ignored how the objectivist methods we were using (employment data trends) formed an important part of what we understand the field of Technical Communication to be. We did not consider how the databases we used and their material-technological frameworks worked to construct the bodies of technical communicators. That we must consider such aspects is exemplified in Barad's (2007) discussion of the famous Stern-Gerlach experiment, where interactions among humans, methods, and material objects produced the phenomena being studied. Without going into specific detail, the experiment results were changed by the fact that one of the assistants smoked cheap cigars. The high sulfur content of those cigars worked to interact with the other chemicals in the experiment to produce results that might otherwise have been invisible.

Thus, the interaction of humans and material objects becomes crucial to what the experiment can show or produce.

As a new materialist method, Nordstrom's object-interview allows the researcher's body and the technologies of production to display their agency in the making of a world, or, as she would call it, a *life*. Using Gilles Deleuze, Nordstrom (2013) shows how "folding subjects and objects help to materialize, or embody, life events in family history" (p. 241). The objects or artifacts of family genealogical history embody the events of an ancestor's life, producing what she calls an object-subject, which is closely linked with an event: "As the event happens in different times and spaces, it defines the folding object-subject again. Each sentence reinvents, or redefines, an object-subject again and again" (p. 242). Each time an object is discussed, held, touched, or otherwise invoked, the object-subject is reinvented again. In Nordstrom's examples of how recording devices are intimate apparatuses of production in qualitative interviews, the object-subject is continually reinvented through interactions with the human bodies of the researchers and genealogists, as well as the technologies of recording. These object-subject-events come together to produce a life, "an atmosphere that surrounds the objects-subjects-events of family history genealogy" (p. 243). "*A life*," Nordstrom continues, "is always already a part of objects-subjects-events and cannot be thought apart from those terms" (p. 243). A life, then, is a mangle of human and nonhuman elements that continually reinvent and reconnect in various ways, and gender matters here.

How, then, do we present the lives of female technical communicators in meaningful ways that demonstrate the challenges they face? One approach could be to perform object-interviews in Technical Communication to see how technological objects change the ways we understand gender and its manifestation among practitioners. Objects and material artifacts, then, do not have to eclipse human being, but they can help human being show up differently. In a way, such an approach can be thought of as a new form of standpoint theory: How is the human being visible from the standpoint of a technological object? How can that object elicit perspectives and notions about the human that would otherwise be invisible?

Future Directions

In the field's focus on the "professionalization" of students and practitioners, we have paid almost no attention to the material-discursive realities of professionals and burgeoning professionals. Patricia Sullivan and Kristen Moore (2013) present us with one option for mentoring female STEM students in

"Time Talk." They "seek ways to make small changes that encourage students to communicate using their professional voices, then help them pause and consider what those voices imply and how they might be reinforced or improved, depending on what they say" (p. 350). To do this, Sullivan and Moore advocate for small changes in the classroom that might help students gain more understanding of how they can control their time and time accounting, which can transfer to internships and workplace environments. Focusing attention on a seemingly minor task like documenting work time is one way, they argue, to affect the material realities of female students. As Sullivan and Moore explain,

> when mentoring female STEM students, . . . we want to encourage them to rethink the ways power can be asserted over them; or, rather, we want to encourage them to acknowledge the power structures that exist and develop tactics for manipulating these structures to benefit their own needs and goals. (p. 345)

Feminist new materialism can extend this work to consider how some of the various material devices in the workplace function to control time. It's not only a matter of teaching students how to account for time; it's also a matter of understanding how material realities and bodies also function to account for and control time. How, for instance, can we bring the idea of biological time to this equation? While Sullivan and Moore acknowledge careers that require practitioners to clock lots of hours in coding or other tasks, they do not consider biological time clocks, which also must be negotiated along with those professional time tasks. Their work, though, is a first step in acknowledging how an exterior agentiality, such as time, produces professional identities and subjectivities.

So, where does this leave us? How can re-seeing the intra-actions of human beings in Technical Communication through feminist new materialism open up a world in which diverse materialities gain rhetorical power? One way involves paying more attention in the classroom to how objects and subjects work to stitch together identities in professional and technical writing. Feminist new materialism would take account of the ways differences that matter accrue in material-discursive apparatuses that construct gender, and it would urge us to take responsibility for these agential cuts in the various methodologies we employ. In simpler terms, what would it mean to have students see technological artifacts as agents rather than as things to be conquered or mastered? Would that give them a different sense of the world? Rather than working against things, could they work with them? And how would such knowledge and experience play out in coworking scenarios or any other kind of distributed work, rife as they may be with gendered power differentials?

I think back to some of the students I have taught who do not see themselves as "experts" of technologies, or who show some reticence with them. Those same students use social media and other digital communications with ease, yet they do not see themselves as capable of learning other digital technologies or advanced software. Perhaps the objectivist approaches to theory in the classroom, the methodologies that the field and university advocate and by which students are taught to do research, reinforce to students that they must "master" the technologies and approaches they use. A feminist new materialist approach in which we teach students to see those technologies as cocreators of the world and as agential elements in any creative situation could provide reserved students with more confidence. It could also allow them to start to see how their own distributed work processes interact with their material realities in ways heretofore unexamined.

In many ways, this chapter argues that we pay more attention to theory and the materials a/effects of our theories on the bodies we teach and impact. Barad (2012) provides the following description of what theory does and what it is meant to do:

> Theorizing, a form of experimenting, is about being in touch. What keeps theories alive and lively is being responsible and responsive to the world's patternings and murmurings. Doing theory requires being open to the world's aliveness, allowing oneself to be lured by curiosity, surprise, and wonder. Theories are not mere metaphysical pronouncements on the world from some presumed position of exteriority. Theories are living and breathing reconfigurings of the world. The world theorizes as well as experiments with itself. Figuring, reconfiguring. (p. 207)

I quote this passage at length to point out that we often associate theory with masculinist discourse that makes pronouncements about the world from an objectivist standpoint. What Barad argues instead is that theory is a form of world-making, an inventional device. It is a heuristic, and our heuristics show us what is and is not possible in the world. In many ways, heuristics are like a scientific apparatus—a material framework through which we make meaning. We have incredible responsibility for what kind of worlds we build via theory. The worlds engendered are human worlds, and the apparatuses we use impact inhabitants, especially in terms of how particular humans and nonhumans come to matter. I see the previous exploration of feminist new materialist approaches as experimenting with the kinds of worlds we can build with and for Technical Communication, worlds that would make gender matter and make the mattering of gender show up for us differently.

Notes

1. According to the Pew Research Center, women, African Americans, and Latinos tend to show higher usage of social media sites. See *The Demographics of Social Media Users* (Duggan & Brenner, 2013).

2. I want to be clear here that men may also be performing these household tasks, but this work has historically been coded as part of a gendered network of caregiving.

References

Barad, K. (2003). Posthumanist performativity: Toward an understanding of how matter comes to matter. *Signs, 28*(3), 801–831.

Barad, K. (2007). *Meeting the universe halfway: Quantum physics and the entanglement of matter and meaning.* Durham, NC: Duke University Press.

Barad, K. (2012). On touching—the inhuman that therefore I am. *Differences, 23*(3), 206–223.

Barnett, S., & Boyle, C. (Eds.). (2016). *Rhetoric, through everyday things.* Tuscaloosa: University of Alabama Press.

Bennett, J. (2009). *Vibrant matter: A political ecology of things.* Durham, NC: Duke University Press.

Coole, D., & Frost, S. (2010). Introducing the new materialisms. In D. Coole & S. Frost (Eds.), *New materialisms: Ontology, agency, and politics* (pp. 1–43). Durham, NC: Duke University Press.

Dolphijn, R., & van der Tuin, I. (2012). *New materialism: Interviews and cartographies.* Ann Arbor, MI: Open Humanities Press.

Duggan, M., & Brenner, J. (2013). *The demographics of social media users, 2012* (Vol. 14). Washington, D.C.: Pew Research Center's Internet and American Life Project.

Durack, K. T. (1997). Gender, technology, and the history of technical communication. *Technical Communication Quarterly, 6*(3), 249–260.

Endres, D., & Senda-Cook, S. (2011). Location matters: The rhetoric of place in protest. *Quarterly Journal of Speech, 97*(3), 257–282.

Gandini, A. (2015). The rise of coworking spaces: A literature review. *Ephemera, 15*(1), 193–205.

Gries, L. E. (2015). *Still life with rhetoric: A new materialist approach for visual rhetorics.* Boulder, CO: Utah State University Press.

Haas, A. M. (2012). Race, rhetoric, and technology: A case study of decolonial technical communication theory, methodology, and pedagogy. *Journal of Business and Technical Communication, 26*(3), 277–310.

Hacking, I. (1988). The participant irrealist at large in the laboratory. *British Journal for the Philosophy of Science, 39*(3), 277–294.

Hallenbeck, S. (2012). User agency, technical communication, and the 19th-century woman bicyclist. *Technical Communication Quarterly, 21*(4), 290–306.

Hekman, S. (2010). *The Material of Knowledge: Feminist Discourses.* Bloomington: Indiana University Press.

Katz, S. B. (1992). The ethic of expediency: Classical rhetoric, technology, and the Holocaust. *College English, 54*(3), 255–275.

Kimball, M. A. (2006). Cars, culture, and tactical technical communication. *Technical Communication Quarterly, 15*(1), 67–86.

Latour, B. (1999). Pandora's hope: Essays on the reality of science studies. Cambridge, MA: Harvard University Press.

Latour, B., & Woolgar, S. (1979). *Laboratory life: The construction of scientific facts.* Beverly Hills, CA: Sage Publications.

Malone, E. A. (2013). Elsie Ray and the founding of STC. *Journal of Technical Writing and Communication, 43*(2), 121–143.

Malone, E. A. (2015). Eleanor McElwee and the formation of IEEE PCS. *Journal of Technical Writing and Communication, 45*(2), 104–133.

McNely, B., Spinuzzi, C., & Teston, C. (2015). Guest editors' introduction: Contemporary research methodologies in technical communication. *Technical Communication Quarterly, 24*(1), 1–13.

Nordstrom, S. N. (2013). Object-interviews: Folding, unfolding, and refolding perceptions of objects. *International Journal of Qualitative Methods, 12*(1), 237–257.

Nordstrom, S. N. (2015). Not so innocent anymore: Making recording devices matter in qualitative interviews. *Qualitative Inquiry, 21*(4), 388–401.

Overmyer, T. (2013). Technical writers and the gender gap: Trending discussions and disparities. *Journal of Purdue Undergraduate Research, 3*, 98–99.

Petersen, E. J. (2014). Redefining the workplace: The professionalization of motherhood through blogging. *Journal of Technical Writing and Communication, 44*(3), 277–296.

Rickert, T. (2013). *Ambient rhetoric: The attunements of rhetorical being.* Pittsburgh, PA: University of Pittsburgh Press.

Ruggles, S., et al. (2010). *Integrated public use microdata series: Version 5.0* [Machine-readable database]. Minneapolis: University of Minnesota.

Russell, D. R. (1997). Rethinking genre in school and society: An activity theory analysis. *Written Communication, 14*(4), 504–554.

Slack, J. D., Miller, D. J., & Doak, J. (1993). The technical communicator as author: Meaning, power, authority. *Journal of Business and Technical Communication, 7*(1), 12–36.

Smith, D. E. (1974). Women's perspective as a radical critique of sociology. *Sociological Inquiry, 44*(1), 7–13.

Spinuzzi, C. (2007). Guest editor's introduction: Technical communication in the age of distributed work. *Technical Communication Quarterly, 16*(3), 265–277.

Spinuzzi, C. (2012). Working alone together: Coworking as emergent collaborative activity. *Journal of Business and Technical Communication, 26*(4), 399–441.

Sullivan, P., & Moore, K. (2013). Time talk: On small changes that enact infrastructural mentoring for undergraduate women in technical fields. *Journal of Technical Writing and Communication, 43*(3), 333–354.

Thompson, I. (1999). Women and feminism in technical communication: A qualitative content analysis of journal articles published in 1989 through 1997. *Journal of Business and Technical Communication, 13*(2), 154–178.

Tuana, N. (2001). Material locations: An interactionist alternative to realism/social constructivism. In N. Tuana & S. Morgan (Eds.), *Engendering rationalities* (pp. 221–243). Bloomington: Indiana University Press.

Tuana, N. (2008). Viscous porosity: Witnessing Katrina. In S. Alaimo & S. Hekman (Eds.), *Material feminisms* (pp. 188–213). Bloomington: Indiana University Press.

5.

How Good Brain Science Gets That Way

Reclaiming the Scientific Study of Sexed and Gendered Brains

Jordynn Jack

In "How Bad Science Stays That Way," Celeste Condit (1996) argues that brain sex research relies on "commonsense understandings of the nature of the dispute about male and female sex and gender" (p. 88), leading to "bad science"—science that produces accounts that are "insufficiently rich to account for the material phenomenon under investigation" (p. 87). Since then, the growth of neuroscience as a discipline has led to more and more accounts that reproduce the tendencies Condit noted, relying on topoi about men's superior map-reading skills or women's interest in shopping. Other researchers in rhetoric have also pointed out these tendencies in brain research. Jeanne Fahnestock (1999), for example, shows how neuroimaging research produces antithetical visual and verbal arguments by exaggerating differences between men and women, using the rhetorical figure of antithesis. In Anne Fausto-Sterling's (2000) terms, this type of research mistakes sex/gender differences for sex *dimorphism*, rhetorically pushing men and women apart by exaggerating what are, in fact, small *average* differences between men and women. More recently, Christa Teston (2016) has shown how brain sex research relies on null hypothesis significance testing, a statistical method that, in Teston's terms, "disciplines scientists to see in certain ways" (p. 49), particularly by rendering as "significant" group differences that may in fact be rather minor (50).

With relation to neuroscience research, in particular, many scholars have thoroughly analyzed and questioned results that purport to identify measurable anatomical or connective differences in male and female brains. Cordelia

Fine's *Delusions of Gender* (2010) and Rebecca Jordan-Young's *Brain Storm* (2010) are both excellent resources for those seeking to understand that research. To date, then, the predominant position among feminist scholars has been to critique brain research, to point out its flawed reasoning, and to lambaste its tendency to uphold antiquated gender stereotypes.

This fallback position, though, poses problems for feminist rhetorical science studies because it begins from a standpoint of critique rather than engagement. Karen Barad (2007), for instance, argues that feminist researchers should avoid the traditional position in science studies, which has been to "position oneself at some remove, to reflect on the nature of scientific practice as a spectator" (p. 247), and, usually, to launch a critique of those practices. Indeed, research in the field of feminist science studies has, since the late 1970s, criticized the masculinist assumptions that have shaped scientific practice—its institutions, language, methods, theories, and norms.

However, scholars have also sought to engage with scientific practice by articulating new forms, including standpoint epistemology (Harding, 2001), situated knowledges (Haraway, 2001), and agential realism (Barad, 2007). While they differ in the particulars, these theories all call attention to how rhetoric constitutes scientific knowledge, the social and cultural are entangled with the material, and feminist researchers should not critique from the sidelines but begin to engage in scientific practice itself. Donna Haraway (2001) writes, for instance, that science "is rhetoric, the persuasion of the relevant social actors that one's manufactured knowledge is a route to a desired form of very objective power" (p. 170); for Haraway, in short, science is a deeply rhetorical endeavor. The goal of feminist science, then, is to use those persuasive tools to offer a

> richer, better account of a world, in order to live in it well and in critical, reflexive relation to our own as well as others' practices of domination and the unequal parts of privilege and oppression that make up all positions. (p. 172)

Barad (2011), similarly, argues for researchers to engage "in practices we call 'science studies' together with practices we call 'science'" (p. 446). As a precursor to this kind of engagement with the neurosciences, I will argue, feminist rhetorical science studies scholars need to reposition themselves, starting not from a place of critique but from an openness to engagement with the brain sciences that does not entail simply embracing or rejecting neuroscience wholesale. This may seem difficult given the tendencies Condit (1996) and Fahnestock (1999) outline. Here, I demonstrate how at least some neuroscience researchers are moving toward a kind of approach more amenable to engagement with humanities researchers. By questioning gender stereotypes and how they emerge in brain research, these scientists open up possibilities for entanglement. In order to identify how alternative kinds of research can be possible, we must turn our attention to the experiment as the unit of analysis. Only by understanding how scientific knowledge is produced through specific experimental entanglements can we begin to identify new possibilities that better capture the interaction of the sociocultural and neurobiological networks through which brain sex/gender differences may (or may not) emerge.

In this chapter, I begin with two psychological studies in order to develop a framework: "Influences of Gender Role and Anxiety on Sex Differences in Temporal Summation of Pain" (Robinson, Wise, Gagnon, Fillingim, & Price, 2004) and "Gender Differences, Motivation, and Empathic Accuracy" (Klein & Hodge, 2001). Next, I show how this framework can be applied to neuroscientific imaging studies, showing in particular how some recent neuroscience research actually supports the viewpoints of feminist researchers who understand the sociocultural and neurobiological to be intertwined. These experiments help to illustrate how sex and gender are not merely preexisting differences that may be revealed through experiments but that sex/gender itself emerges through experiments, appearing differently depending on the experimental apparatus used.

On Neurorhetorics

This essay develops out of the area of study referred to as "neurorhetorics," or the study of "how discourses about the brain construct neurological difference, how to operationalize rhetorical inquiry into neuroscience in meaningful ways, and what those constructions imply for contemporary public discourse" (Jack, 2010, p. 406). Within this area of study, scholars might explore both the rhetoric of neuroscience, or how neuroscientific knowledge circulates through discourse, and the neuroscience of rhetoric, or how insights from neuroscience might inform rhetorical theory. Ideally, as I have argued elsewhere with L. Gregory Appelbaum, these types of study should be intertwined (Jack & Appelbaum, 2010). In other words, in order to understand how neuroscience concepts can contribute to rhetorical theory, I maintain that one should understand how those concepts are themselves established rhetorically. Methodologically, this means paying careful attention to scientific knowledge in the genre in which it circulates, the scientific article; examining the methods, tools, and concepts used in those articles; and remaining cautious when extending the implications of that research outward to rhetoric.

This research participates in the broader call for humanistic and social scientific approaches to neuroscience put forward by scholars such as Suparna Choudhury, Saskia Kathi Nagel, and Jan Slaby (2009); Des Fitzgerald and Felicity Callard (2015); and others. Among those scholars, I find Fitzgerald and Callard's (2015) call for a return to the experiment particularly useful. They have argued that *"experimental labour itself"* is worth "sustained attention from social scientists and humanists" (p. 5), since it offers a way not only to better understand how neuroscience knowledge is produced but also to engage with neuroscience and even participate in it (and here they draw inspiration from Barad's [2007] call for an "entanglement" with the natural sciences [p. 20]).

Before moving on, it is important to recognize how neuroscience functions rhetorically. To date, much of the scholarship on the rhetoric of neuroscience has focused on public discourse. Scholars have examined how neuroscience involving neuroimaging tools such as functional magnetic resonance imaging (fMRI) plays out in the popular sphere. As Joseph Dumit (2004) has explained, these imaging tools are constituted rhetorically as scientifically valid measurement tools, but they also "make claims on us" (in Dumit's words), establishing different types of people in public rhetorics about the "depressed brain" or the "creative brain," for instance (p. 5). Neuroscientists have themselves called attention to the problem of popularization, suggesting that neuroscience explanations hold a "seductive allure" that belies the tentativeness of current research findings (Weisberg, Keil, Goodstein, Rawson, & Gray, 2008).

My focus here, though, is on the rhetoric of scientific research articles and of experiments themselves in neuroscience. One important point to mention is that neuroscience is itself an interdiscipline, one that draws on insights from fields such as psychology, biology, computing, physics, engineering, and linguistics. In the area of cognitive neuroscience, in particular, much of the rhetorical work required involves bridging psychology and neurobiology, or linking psychology concepts such as memory, attention, and reward to neurological locations or substrates. This rhetorical work begins long before an experimental result is published; indeed, the research questions asked, the experimental protocols designed, and the apparatus used (including everything from psychometric questionnaires to measurement tools such as fMRI) involve rhetorical decisions. As Barad (2007) puts it,

> every measurement involves a particular choice of apparatus, providing the conditions necessary to give meaning to a particular set of variables, at

the exclusion of other essential variables, thereby placing a particular embodied cut delineating the object from the agencies of observation. (p. 115)

In neuroscience research on sex/gender differences, one choice is to assume such variances exist a priori (outside of scientific observation) and to then devise an experiment that demonstrates the extent of those differences. A different approach, however, is to assume that such variances are constituted in and through experiments, and to therefore design an apparatus that manipulates those differences themselves. In the latter, sex/gender take on other meanings depending on the apparatus.

Gender-Aware Methodologies

Researchers interested in sex/gender differences in neuroscience often begin from an antithetical commonplace about men and women (such as the idea that men are more interested in sports than women are) that is taken as objective and real, and then seek to explain it by blending psychological and neuroscientific methods. For instance, researchers might have measured spatial ability in male and female participants using an object rotation task while they were in an fMRI machine; the researchers would then determine, on average, how well men and women performed the task and identify which brain regions men and women used while doing it using null hypothesis significance testing, as Teston has outlined. The starting assumption they made—that men and women have different spatial reasoning abilities—shapes the materials of the experiment itself, including the questionnaire used, the brain regions examined in the fMRI machine, the use of the machine itself, and so on. This approach parallels the assumptions Barad (2007) outlines as the classical epistemological and ontological assumptions underlying Newtonian physics, which assumes, first, "that the world is composed of individual objects with individually determinate boundaries and properties whose well-defined values can be represented by abstract universal concepts that have determinate meanings independent of the specifics of the experimental practice" and, second, that it is possible to measure those objects neutrally, as separate from "the agencies of observation" (p. 107). In the case of neuroscience, then, researchers operating under this set of assumptions believe that sex and gender are clearly delineated objects that exist "out there" in the world: this includes "men" and "women" as well as "male" and "female" brains. It is this type of research that Condit critiqued in her 2008 study for its oversimplistic tendency to reify the very objects it sought to investigate.

However, researchers who make a different rhetorical decision might design a study another way. A feminist perspective, for instance, would require researchers, in Condit's (2008) words, to "ask questions such as 'under what conditions do people manifest particular behavioral patterns out of their available repertoires?' rather than simply asserting that males and females have fixed behavioral patterns" (p. 493). Here, I consider two studies to illustrate how starting from a different assumption leads researchers to pay attention to different aspects of the experiment and to recognize the entanglement of societal gender norms and priming effects with experimental apparatus and results. These studies demonstrate, in Barad's (2007) terms, how "measurement practices are an ineliminable part of the results obtained" (p. 121). By selecting different ways of measuring pain and empathy, in these instances, researchers allow sex/gender differences to materialize in various ways.

Typically, studies of pain call on participants to rank their pain levels on a scale; most studies have shown that men have higher pain tolerances than women. These studies assume that men and women exist outside of the experimental apparatus, as do pain tolerances; experiments, then, must simply measure those differences. However, in Robinson et al.'s (2004) study, "Influences of Gender Role and Anxiety on Sex Differences in Temporal Summation of Pain," researchers demonstrated how those differences in fact are produced by experimental apparatuses, including different scales of measurement, and how they are deployed within an experimental setting. In this study, participants first completed a set of questionnaires dealing with how they experienced pain, including one called the "Gender Role Expectations of Pain Questionnaire." This questionnaire asks subjects to report how the typical man and woman responds to pain, and then asks participants to rate themselves. One question asks how willing the typical man or woman is to report pain. In other words, this question asks participants whether they have absorbed the common social belief that men are less likely to admit to feeling pain than women are. Next, participants completed several types of tests in which they were exposed to physical pain via heat pulses.

Robinson et al. found that, while women reported greater levels of pain during these experiments, those differences corresponded to the participants' responses on the questionnaires. In particular, participants who believed themselves (and their gender) more willing to report pain experienced more pain. The authors concluded that pain is "a hypothetical psychologic construct" (p. 80), not simply a physical reality. To put this another way, we might say that these results provide an example of Barad's theory of entanglement—the material, embodied pain response is entangled with cultural expectations.[1]

This experiment further illustrates, in Barad's (2007) terms, how theoretical concepts, such as pain as well as sex/gender, "are not ideational in character but rather *specific physical arrangements*" (p. 139). Here, the physical arrangements include study participants, the questionnaires they complete, the researchers themselves, and the specific way pain was measured (via heat pulses) as well as the ways sex/gender were defined, including in the recruitment of subject participants assumed to be unproblematically male or female prior to the study itself. By setting up their apparatus differently—simply by using a different measurement scale than was used in other studies—researchers materialized pain and sex/gender differently from other researchers who assumed both objects existed independently from and prior to the study itself.

In the next example, Klein and Hodges's (2001) study "Gender Differences, Motivation, and Empathic Accuracy," researchers found that material elements of the study produced a significant (but not intentional) experimental effect: making gender differences appear or disappear in the study of empathy. Klein and Hodges sought to investigate an intriguing problem in the study of "empathic accuracy," or the ability of participants to infer accurately the thoughts and feelings of others. To judge empathic accuracy, one common method has been to have participants view videos of an individual discussing an event or problem. The participants must judge how the individual felt at key moments in the video (as determined by the individuals in the video, who noted their emotions or thoughts after watching their videos). A striking issue has emerged from studies using this method. Namely, researchers found no gender differences in empathic accuracy in the first seven studies. However, when they changed the written form on which participants recorded their findings, a gender difference began to emerge, with women performing better than men. On the new form, participants were asked to indicate how accurately they thought they had judged emotional states for each key moment in the video. In other words, the new forms made participants conscious of empathic ability—a trait commonly linked to women—and, in so doing, may have motivated women to do better at the task.

To test this hypothesis, Klein and Hodges performed two different experiments, using male and female college students as participants. In the first experiment, participants watched a video of a female student relating an academic problem. Half of the participants then completed a questionnaire reporting how sympathetic they felt, and then completed the empathic accuracy task. For the other half of the participants, the order was reversed. Presumably, the sympathy questionnaire would evoke gendered stereotypes, priming women to perform better on the test of empathic accuracy. The experimenters also manipulated the instructions accompanying the empathy test. For one third of the participants, the task was introduced as a test of cognitive ability; in the second third, it was described as a test of empathy; and in the final third, no mental ability was invoked.

Klein and Hodges found that women performed better on the empathic accuracy test when they first completed the sympathy test. They also performed better than men on the empathic accuracy test when the instructions invoked the concept of empathy, but not in the control condition or when the test evoked cognitive ability.

In the second experiment, Klein and Hodges used the same protocol, only this time the test subjects were paid for each correct answer on the empathic accuracy test. (In the control condition, no payment was received; a third group received feedback on their performance but no money.) When paid for the accuracy of their responses, men and women scored equally well on the empathic accuracy test. The authors concluded that

> if a woman is aware that the task she is completing is assessing her empathic capabilities, it may be important for her to perform well. She therefore may be more successful than a man completing the same objective measurement of empathy because of her increased level of motivation. (p. 721)

In short, Klein and Hodges determined that a minor shift in the experimental apparatus could significantly influence results, in effect erasing gender differences that had been widely upheld in other scientific studies and that had become a cliché in contemporary discourse about gender differences.

Klein and Hodges were thus able to identify elements in the material-discursive entanglement of experimental apparatus and gender norms. For one, the form used in a study could prime a gendered concept in participants (the idea that women are more empathetic than men), leading to results that seem to uphold a gender binary. Alternatively, introducing a different material-discursive element into the study (money) could produce a different outcome, in effect erasing gender differences by motivating men to perform better on the test.

Studies such as the two I have examined here have been taken up by other researchers seeking to understand the social and rhetorical contexts that produce gender differences. Moreover, researchers are questioning the concepts themselves that tend to be associated with such research—pain and empathy are two of the concepts most commonly linked to gender. The Robinson et al. (2004) study has been cited in a number of review articles and meta-analyses that now suggest a consensus: namely, that gender roles shape the experience of pain, with those who identify as more masculine reporting lower sensitivity to pain. Rather than taking this as evidence of some innate connection between masculinity and inherent pain tolerance, researchers concluded that "learned masculinity encourages stoicism" as well as "displays of withstanding pain." (Alabas, Tashani, Tabasam, & Johnson, 2012). Similarly, psychologist Andrea Lobb (2013) argues that the concept of empathy is "laden with dubious assumptions about sexual difference" (p. 427). More specifically, Lobb claims that the concept of empathy allows "regressive sex stereotypes" to be worked back into psychology research via the "moral cover" of empathy (p. 427). Klein and Hodges's (2001) study forms part of the centerpiece of Lobb's argument, namely, that the female advantage in empathy is not itself scientifically proven. From there, Lobb (2013) goes on to question the moral and ethical implications of assigning as important a value as empathy to just one sex. In other words, at least some researchers are beginning to recognize the entanglement of sociocultural and neurobiological factors that shape pain or empathy. However, these studies still seem to have swapped the idea of sex/gender differences for that of stereotypes; in other words, while they now question that sex/gender differences exist neatly outside of the experiments that purport to measure them, they still assume that gender stereotypes are obvious, preexistent concepts.

Together, the studies I have considered so far suggest two ways that neuroscience experiments can constitute or challenge sex/gender difference. First, sociocultural expectations may prefigure experiments, insofar as participants implicitly evoke their own beliefs (such as their belief that women are less tolerant of pain) within an experimental setting. Second, material elements of the experimental situation itself may shape responses; even small items like a form used as part of an experimental protocol may prime gendered responses. Other elements, such as rewards given for participation, may also shape results. As feminist scholars of rhetoric and science, we should be careful not to overlook these elements when interpreting an experimental study or drawing from its results. Not only can these small differences influence results, but they can also do so in ways that either force apart sex/gender distinctions or bring them closer together. As experts in rhetoric, we can be effective research partners in neuroscientific studies, to the extent that we are trained to identify the possible means of persuasion, even within a scientific experiment.

In the next section, I consider whether and how neuroimaging experiments investigating sex/gender differences might draw on these perspectives to demonstrate the entanglement of rhetorical and cultural gender expectations with the materiality of the brain.

Neuroimaging Sex and Gender

Despite the promising frameworks generated by the previously discussed studies, neuroscience research often starts from the assumption that sex/gender differences exist and seeks to identify those variances, rather than asking the question of whether observable sex/gender differences are in fact artifacts of cultural beliefs and rhetorical contexts of the experiment itself. To determine the extent to which the former approach dominates, I pulled all abstracts in which *sex* or *gender* appeared from a database of neuroimaging studies that appeared between 2004 and 2010 in five neuroscience journals.[2] Working with my research assistant, Jennifer Stockwell, we then eliminated abstracts referring to studies that did not explicitly examine sex/gender differences; these included forty abstracts that simply listed the gender of the participants and those that used terms such as *engender*. This left a list of forty-nine studies.

Jennifer and I then coded those studies, separately, identifying the hypotheses and study designs as either demonstrating sex/gender differences (D); studying phenomena specifically in either men's or women's brains (G); or questioning sex/gender differences themselves (Q). (Agreement between the research assistant and I neared 95 percent). Overall, thirty-one studies fell into the (D) category, fourteen into the (G) study, and only four into the (Q) category. These results suggest that opportunities for engagement with the neurosciences are plentiful. Heeding Fitzgerald and Callard's (2015) call for humanities and social science researchers to work alongside neuroscientists, rather than retreating to the more comfortable position of armchair critique, could allow us to enrich those studies that continue to start from the assumption of difference rather than questioning whether, why, and in what conditions it exists.

First, let's consider the studies that seek to identify sex/gender differences in the brain, or the (D) studies. While a thorough review of these studies is beyond the scope of this chapter, in general, we find that the frameworks Fausto-Sterling (2000), Condit (1996), and Fahnestock (1999) problematized persist in neuroscience research. These kinds of studies often *start* with an antithesis, posing a stereotype as common knowledge readily observable in day-to-day life, and then asking whether it has a neurological basis.

Often, these antitheses and generalizations appear in the first paragraph of the research article. Consider the following examples:

- "Behavioral studies suggest that females often perform better in emotional tasks than males." (Schulte-Rüther, Markowitsch, Shah, Fink, & Piefke, 2008, p. 393)
- "In the past few decades, scientists and the public alike have debated the existence of gender differences in mathematical skills." (Keller & Menon, 2009, p. 342)
- "Men and women ... differ markedly in aspects of sexual behavior, such as the reportedly greater male interest in and response to sexually arousing visual stimuli." (Hamann, Herman, Nolan, & Wallen, 2004, p. 411)
- "Men outperform women on several spatial ability measures." (Seurinck, Vingerhoets, de Lange, & Achten, 2004, p. 1440)
- Disorders such as depression and post-traumatic stress disorder "have a substantially greater prevalence (1.5–3 times) in females than males." (Williams et al., 2005, p. 618)

We might summarize the arguments in these first lines as follows:

- Women are more empathetic than men.
- Men are better at math than women.
- Men like porn.
- Women can't read maps.
- Women are more emotional than men.

While some of these studies couch those stereotypes in results from behavioral or psychometric research, they nonetheless take a stereotype or a generalization at face value and then seek to identify its neural correlates, rather than seeking to identify where it comes from or whether it reflects rhetorical and cultural entanglements like the ones outlined in the previous section. These studies, then, assume that gender stereotypes are somehow outside of scientific practices, ignoring how "the social and the scientific are co-constituted" (Barad, 2007, p. 168). In fact, these studies do not reflect an existing reality about sex/gender and the brain but in fact contribute to the production of sex/gender differences both within the scientific experiment itself and in the discourses surrounding them; the prominence of these types of experiments in popular sciences news headlines offers one example of this.

The (G) studies also seem to reflect gendered stereotypes. These studies use gender stereotypes or norms as the basis for selecting participants, since they typically choose to study individuals of only one sex. In many cases, no rationale is given for deciding to study only one group; the reader is left to fill in the assumptions underlying the choice. For instance, one study of pedophiles used only male participants who were attracted to female children but did not explain this choice (Schiffer et al., 2008). This pick of study subject and study design assumes the common belief that pedophiles are men who prey on young girls. In others, a weak rationale was given for confining a study to only one sex, as was the case in a study of responses to sexual stimuli that used only men viewing pornographic images of women; the authors assumed that men were naturally more responsive to such images than women (Borg, Lieberman, & Kiehl, 2008, p. 1531). Finally, some of the studies used only men or women because they were studying sexual responses unique to male or female anatomy. Here, the reasoning seems to be that men and women have unique neuroanatomy due to their different anatomical structures, so men and women are best studied separately. These studies are also typically confined to either heterosexual or homosexual participants. One study of sexual response, for instance, involved heterosexual males and their female partners, who participated by stimulating their boyfriends (Georgiadis et al., 2010). All these approaches, then, take sex/gender differences as preexisting and natural, with the study merely revealing those differences. Nonetheless, we can see even from these schematic descriptions that each study materializes sex/gender/sexuality differently through its choices. By focusing on men's responses to visual stimuli, for instance, Jana Schaich Borg, Debra Lieberman, and Kent A. Kiehl (2008) constitute men as sexually responsive to pornography and seek to produce such responses in the brain.

Next, let's examine the four studies (Q) that sought to question sex/gender stereotypes. Two of these studies (Mitchell, Ames, Jenkins, & Banaji, 2009; Quadflieg et al., 2009) were framed as research into the phenomenon of social stereotyping, with gender offering a case study with which to explore that phenomenon. In other words, these studies are not framed explicitly as inquiries into whether gender stereotypes have some neurological basis; instead, they seek to understand how those stereotypes work in the first place.

Jason Mitchell et al.'s "Neural Correlates of Stereotype Application" (2009) begins by outlining the broad phenomenon by which people mentalize about others by activating stereotypes. The researchers explain that stereotypes guide people's judgments about others' psychological characteristics and behaviors (p. 594). For example, the researchers showed participants a picture of a man or woman and then asked them to judge whether or not a statement would apply to that person, such as "likes scented candles" or "enjoys watching football" (p. 595). In rhetorical terms, the researchers prompted people with commonplaces or topoi about what men and women are supposed to like.

The authors suggest that we may draw on the right prefrontal cortex, an area of the brain associated with categorization, when making stereotyped judgments. Those participants who scored higher on measures of gendered attitudes (those who had been exposed to more gendered topoi and persuaded by them), in fact, showed greater activations in that region of the brain. The authors concluded that when applying gendered stereotypes to individuals, we draw on a more fundamental cognitive process of categorization. In Burkean terms, we might term this a sense of "what properly goes with what," or a piety that develops over time through repeated exposure to gendered topoi (1984, p. 74). This research, then, points to a potential example of what Fausto-Sterling (2000) and Barad (2007) term "dynamic systems" or "entanglements," respectively, or the mutual imbrication of neurological functions with social and rhetorical constructs of sex/gender. At the same time, however, it is important not to overlook how invoking gender stereotypes in this study itself materializes and cements those stereotypes, contributing to the very phenomenon it hopes to explain.

We see similar support for a neurological basis for gender stereotypes in a study by Susanne Quadflieg et al., "Exploring the Neural Correlates of Social Stereotyping" (2009). In this study, participants were asked to indicate whether certain kinds of activities, such as mowing the lawn, watching talk shows, or taking photographs, corresponded more with men or women (or neither). As they were doing this task, researchers monitored participants' blood oxygenation activation in an fMRI machine. The researchers correlated those activation responses with participants' results on two questionnaires, the "Implicit Association Test" and the "Attitudes toward Women Scale." They found patterns of activation in regions of the brain associated with activity knowledge and with evaluative or emotional processing (the amygdala). An interesting conclusion from this study is that gender stereotyping is fundamentally evaluative (p. 1567), but that it also depends on learned associations. In fact, those who scored highest on the scales used to test gendered beliefs had the greatest activation in the right amygdala, a region of the brain associated with "stimuli that have acquired emotional significance through learning rather than based on some innate propensity" (p. 1567). The authors concluded that while stereotyping is a nearly universal human capacity, "activity in the neural circuitry supporting these responses is sensitive to the strength with which gender-based beliefs are endorsed in everyday life" (p. 1567). This is a provocative finding. It suggests that rather than viewing sex/gender differences as hardwired or innate (as many seem to do), scientists might now consider those differences to be based on a fundamental cognitive capacity to make judgments, coupled with a learned capacity to make those judgments based on previous exposure to gendered topoi. This study contends that the neurobiological and sociocultural are clearly entangled, with repeated exposure to such stereotypes producing material differences in the brain. Yet Barad would remind us not to view this study (or any other experimental study) as simply reflecting a reality about gender stereotypes; as with the studies of pain and empathy earlier, we must recognize the experimental apparatus itself as constitutive of this understanding of the phenomenon of gender stereotypes. In other words, this study constitutes gender stereotypes as binaries (male/female, masculine/feminine) that we can associate with specific tasks such as mowing the lawn; individuals differ only by the extent to which they put stock in those stereotypes.

Two other studies take the very act of ascribing sex/gender to individuals as the object of study. In the first example, Dutch researchers investigated how individuals make judgments about an unknown or invisible speaker's age, sex, and social background by interpreting their voice. To examine this phenomenon, the researchers recorded sentences pronounced by male or female speakers, children or adults, and, in the final condition, speakers with an accent associated with either an upper-class or lower-class accent (in Dutch). Some of these sentences conveyed information that would be expected of the speaker, and others did not. For instance, the incongruent utterances included a male speaker saying, "My favorite colors are *pink* and lime green" or a child saying, "Every evening I drink a glass of *wine* before going to bed" (Tesink et al., 2009, p. 2086). These sentences violated common social stereotypes.

While the authors do not speculate about what their findings mean with respect to gender stereotyping (or other forms of stereotyping) in particular, they did find activations in the brain when incongruent sentences were uttered that suggested "increased and more prolonged efforts to search and retrieve semantic knowledge about the likelihood of events occurring in the real world" (p. 2097). These activations occurred in the left inferior frontal cortex, an area of the brain associated with language comprehension. The authors further suggest that the process of "unification" also entails other kinds of information—not just words, but gestures and other extralinguistic vectors of information. Thus, this unifying process involves a variety of inputs, linguistic and otherwise, where individuals try to square incongruent information with their existing assumptions about sex and gender.

A final study in our sample examined a similar process of gender identification, but this time in relation to human motion—specifically gait. The study asked how it is that we ascribe gender to an individual's walk. To answer this question, the researchers attached lights to the major joints of an actor's body. The actor then portrayed male, female, or gender-ambiguous walks, and participants viewing only those lights were asked to identify the gender of the walker. Importantly, the researchers found that this discrimination of gender is contextual. That is, after viewing a "female" walk, viewers tended to identify the next stimulus as a "male" walk, and vice versa. The authors concluded that "gender identification of human walkers is rapidly malleable and subject to adaptation" (Jordan, Fallah, & Stoner, 2006, p. 739). This study, then, perhaps goes the furthest of the ones reviewed here in recognizing gender as itself constituted through the experiment rather than existing a priori in the form of gendered bodies or stereotypes.

Together, this group of studies suggests that researchers who seek to understand how it is that we ascribe gender differences to others can help depolarize male and female brains by refusing the antithetical reasoning foundational to so many other studies. They also begin to resist the polarization between sociocultural and neurobiological elements, showing how they are also entwined. While by no means the most common approach among the studies we surveyed, this last group does indicate that feminist scholars in the rhetoric of science need not eschew brain research completely, and can even find moments of possibility within the experimental sites given here. Nonetheless, feminist scholars might push these researchers further to recognize gender-in-the-making within experiments themselves. For example, the voice study used voices that were presumably easily identified as male or female; while the authors do not describe how they chose such voices, we can assume that the "male" voices were deeper and the "female" voices higher. This marks an assumption that gendered voices match straightforwardly with gendered bodies. This study, then constitutes gender as something we can determine from the tone and timber of one's voice. In other words, vestiges of what Barad refers to as "representationalism" (the assumption, in this case, that "men" and "women" and "male voice" or "female voice" exist "out there" somewhere) continue to shape these studies, even when they seek to explain how sex and gender function neurologically rather than taking them as givens.

Conclusion

Good brain science can emerge when researchers recognize that psychological concepts, experimental protocols, and apparatus materialize sex/gender, rather than simply reflecting or explaining it. The articles reviewed here take us some of the way toward that goal. The psychology studies I've highlighted have helped to shift the conversation within psychology that takes men's greater tolerance for pain, or women's greater empathic accuracy, as natural, as givens. Further, the neuroscience studies help to suggest how the very act of ascribing gender to another person depends on neurological operations as well as social and cultural inputs. My point here, though, is not simply to suggest that the specific results of these research articles necessarily improve on what feminist rhetoricians already know about sex and gender. Instead, I hope this analysis demonstrates how and why feminist rhetoricians should pay close attention to the rhetoricity of science at the level of the research article or experiment so that we may better understand how the decisions scientists make may or may not lead toward a feminist biology.

While I find the articles included here promising in that regard, they do have limitations. For one, the studies examined drew from relatively dated feminist theories. While these studies help to investigate the extent to which gender differences are socially and rhetorically produced, both within a culture and within an experimental setting, few studies take up a perspective of gender as polymorphic or diverse. Most studies assumed participants were naturally male or female, masculine or feminine. The study designs did not allow for a more complex model that would include gender diversity—that is, a wider range of sexed and gendered identities, including transgendered, genderqueer, or agender. That perspective might allow us the possibility to make even richer identity arguments.

In addition, the studies examined did not take an intersectional approach, which would consider gender alongside the dynamics of race, gender, class, sexuality, nationality, and disability. Most of the studies reviewed recruited participants mainly from undergraduate psychology courses—a common practice in psychological research that nonetheless risks generalizing heavily from a specific group of students and ignores the potential differences between individual brains across time, place, and culture. For instance, the types of stereotypes questioned in these studies might reflect mainstream white, middle-class stereotypes (that men don't like shopping, for example, or that women like scented candles, are more empathetic than men, and so on). These kinds of assumptions do not necessarily persist across different cultures, or even across time within a given culture.

Nevertheless, the examples given here suggest that feminist researchers in the rhetoric of science need not dismiss all brain research as sexist. In fact, because they demonstrate how sex-stereotyped behaviors can be constituted differently through minute changes (such as the type of form given), the scientific studies in this chapter offer useful thought experiments for thinking through sex/gender as a phenomenon more broadly—how material and discursive interactions make and remake gender continually and how science contributes to those makings. After all, as Barad (2007) writes, "there isn't one set of material practices that makes science, and another disjunct set that makes social relations; one kind of matter on the inside, and another on the outside. The social and the scientific are co-constituted" (p. 168). This less deterministic view of gendered and sexed brains answers Condit's (2008) call for feminists to "grapple with a more complex account of biology," which she describes "as treating biological organisms as belonging to non-homogenous sets that change through time and whose characteristics are a product of the variable environments (or 'cultures') through which they move" (p. 493). I have argued that, for rhetoricians, this project requires a renewed impetus not only to analyze and disrupt stereotypical portrayals of men and women in scientific research but also a willingness to engage with brain science at the level of the experiment and research article.

Notes

1. For a more in-depth analysis of how pain, in particular, materializes in scientific experiments and discourses, consult S. Scott Graham's (2015) *The Politics of Pain Medicine: A Rhetorical-Ontological Inquiry.*

2. For study methodology, see "Mapping the Semantic Structure of Cognitive Neuroscience" (Beam, Applebaum, Jack, Moody, & Huettel, 2014).

References

Alabas, O. A., Tashani, O. A, Tabasam, G., & Johnson, M. I. (2012). Gender role affects experimental pain responses: A systematic review with meta-analysis. *European Journal of Pain, 16*(9), 1211–1223.

Barad, K. (2007). *Meeting the universe halfway: Quantum physics and the entanglement of matter and meaning.* Durham, NC: Duke University Press.

Barad, K. (2011). Erasers and erasures: Pinch's unfortunate "uncertainty principle." *Social Studies of Science, 41*(3), 443–454.

Beam, E., Appelbaum, L. G., Jack, J., Moody, J., & Huettel, S. A. (2014). Mapping the semantic structure of cognitive neuroscience. *Journal of Cognitive Neuroscience, 26*(9), 1949–1965.

Borg, J. S., Lieberman, D., & Kiehl, K. A. (2008). Infection, incest, and iniquity: Investigating the neural correlates of disgust and morality. *Journal of Cognitive Neuroscience 20*(9), 1529–46.

Burke, K. (1984). *Permanence and change: An anatomy of purpose* (3rd ed.). Berkeley: University of California Press.

Choudhury, S., Nagel, S. K., & Slaby, J. (2009). Critical neuroscience: Linking neuroscience and society through critical practice. *BioSocieties, 4*, 61–77.

Condit, C. (1996). How bad science stays that way: Brain sex, demarcation, and the status of truth in the rhetoric of science. *Rhetoric Society Quarterly, 26*(4), 83–109.

Condit, C. M. (2008). Feminist biologies: Revising feminist strategies and biological science. *Sex Roles, 59*(7), 492–503.

Dumit, J. (2004). *Picturing personhood: Brain scans and biomedical identity.* Princeton, NJ: Princeton University Press.

Fahnestock, J. (1999). *Rhetorical figures in science.* New York, NY: Oxford University Press.

Fausto-Sterling, A. (2000). *Sexing the body: Gender politics and the construction of sexuality.* New York, NY: Basic Books.

Fine, C. (2010). *Delusions of gender: How our minds, society, and neurosexism create difference.* New York, NY: W. W. Norton.

Fitzgerald, D., & Callard, F. (2015). Social science and neuroscience beyond interdisciplinarity: Experimental entanglements. *Theory, Culture, and Society, 32*(1), 3–32.

Georgiadis, J. R., et al. (2010). Dynamic subcortical blood flow during male sexual activity with ecological validity: A perfusion fMRI study. *NeuroImage, 50*(1), 208–216.

Graham, S. S. (2015). *The politics of pain medicine: A rhetorical-ontological inquiry.* Chicago, IL: University of Chicago Press.

Hamann, S., Herman, R. A., Nolan, C. L., & Wallen, K. (2004). Men and women differ in amygdala response to visual sexual stimuli. *Nature Neuroscience, 7*(4), 411–416.

Haraway, D. (2001). Situated knowledges: The science question in feminism and the privilege of partial perspective. In M. Lederman & I. Bartsch (Eds.), *The gender and science reader* (pp. 169–188). London, UK: Routledge.

Harding, S. (2001). Feminist standpoint epistemology. In M. Lederman & I. Bartsch (Eds.), *The gender and science reader* (pp. 145–168). London, UK: Routledge.

Jack, J. (2010). What are neurorhetorics? *Rhetoric Society Quarterly, 40*(5), 405–410.

Jack, J., & Appelbaum, L. G. (2010). "This is your brain on rhetoric": Research directions for neurorhetorics. *Rhetoric Society Quarterly, 40*(5), 411–437.

Jordan, H., Fallah, M., & Stoner, G. R. (2006). Adaptation of gender derived from biological motion. *Nature Neuroscience, 9*(6), 738–739.

Jordan-Young, R. M. (2010). *Brain storm: The flaws in the science of sex differences.* Cambridge, MA: Harvard University Press.

Keller, K., & Menon, V. (2009). Gender differences in the functional and structural neuroanatomy of mathematical cognition. *NeuroImage, 47*(1), 342–352.

Klein, K. J. K., & Hodges, S. D. (2001). Gender differences, motivation, and empathic accuracy: When it pays to understand. *Personality and Social Psychology Bulletin, 27*(6), 720–730.

Lobb, A. (2013). The agony and the empathy: The ambivalence of empathy in feminist psychology. *Feminism and Psychology, 23*(4), 426–441.

Mitchell, J. P., Ames, D. L, Jenkins, A. C., & Banaji, M. R. (2009). Neural correlates of stereotype application. *Journal of Cognitive Neuroscience, 21*(3), 594–604.

Quadflieg, S., et al. (2009). Exploring the neural correlates of social stereotyping. *Journal of Cognitive Neuroscience, 21*(8), 1560–1570.

Robinson, M. E., Wise, E. A., Gagnon, C., Fillingim, R. B., & Price, D. D. (2004). Influences of gender role and anxiety on sex differences in temporal summation of pain. *Journal of Pain, 5*(2), 77–82.

Schiffer, B., et al. (2008). Functional brain correlates of heterosexual paedophilia. *NeuroImage, 41*(1), 80–91.

Schulte-Rüther, M., Markowitsch, H. J., Shah, N. J., Fink, G. R., & Piefke, M. (2008). Gender differences in brain networks supporting empathy. *NeuroImage, 42*(1), 393–403.

Seurinck, R., Vingerhoets, G., de Lange, F. P., & Achten, E. (2004). Does egocentric mental rotation elicit sex differences? *NeuroImage, 23*(4), 1440–1449.

Tesink, C. M., et al. (2009). Unification of speaker and meaning in language comprehension: An fMRI study. *Journal of Cognitive Neuroscience, 21*(11), 2085–2099.

Teston, C. (2016). Rendering and reifying brain sex science. In S. Barnett & C. Boyle (Eds.), *Rhetoric, through everyday things* (pp. 42–54). Tuscaloosa: University of Alabama Press.

Weisberg, D. S., Keil, F. C., Goodstein, J., Rawson, E., & Gray, J. R. (2008). The seductive allure of neuroscience explanations. *Journal of Cognitive Neuroscience, 20*(3), 470–477.

Williams, L. M., et al. (2005). Distinct amygdala-autonomic arousal profiles in response to fear signals in healthy males and females. *NeuroImage, 28*(3), 618–626.

6.

Representing without Representation

A Feminist New Materialist Exploration of Federal Pharmaceutical Policy

Daniel J. Card, Molly M. Kessler, and S. Scott Graham

Recent scholarship in science and technology studies (STS), political science, media studies, political theory, historiography, and rhetoric is actively embracing ontological inquiry.[1] Although this diverse body of scholarship styles itself under a wide array of brands from speculative realism and object-oriented ontologies to alien phenomenology and multiple ontologies, each variant is ultimately committed to a handful of central points. Thus Diana Coole and Samantha Frost (2010) offer "new materialisms" (NM) as a convenient catchall term for these recent intellectual efforts devoted to making objects, things, matter, and the concrete central to inquiry. In addition to this sustained attention to objects, NM typically comes with an explicit rejection of the postmodern preoccupation with the social/cultural/discursive constitution of reality.[2] Correspondingly, much NM scholarship often positions itself as fundamentally opposed to inquiry into epistemology and representation.[3]

The adoption of NM in feminist STS and feminist rhetorical science studies represents a significant shift in theories and modes of inquiry. Indeed, writing on this shift, Susan Hekman (2008) ultimately concludes that the postmodern project has failed. As she writes, "instead of deconstructing the discourse/reality dichotomy, instead of constructing a new paradigm for feminism that integrates the discursive and the material, feminism has instead turned to the discursive pole of the discourse/reality dichotomy" (p. 86). This concern over the status of postmodern intellectual efforts pervades NM inquiry. Bruno Latour (1993) famously declared, "Postmodernism is a symptom, not a fresh solution. It lives under the modern Constitution, but no longer believes in the guarantees the Constitution offers" (p. 46). The modern constitution here, which refers to, among other things, dichotomies between subject and object or nature and culture, remains a problem in a postmodern world.

Much in the vein of Latour's argument in *We Have Never Been Modern* (1993), NM philosopher Levi Bryant (2011) argues:

> As a consequence of the two world schema, the question of the object, of what substances are, is subtly transformed into the question of how and whether we know objects. The question of objects becomes the question of a particular relationship between humans and objects. This, in turn, becomes questions of whether or not our representations map onto reality. (p. 16)

Here, two worlds—one *material* and one *representational*—are posited in both the modernist and postmodernist position. While modernists were concerned with questions of objects or substances, postmodernists rejected the notion these could be known, and focused instead on *representations*. That is, postmodern scholarship denied all access to an independent reality. All materiality was reduced to representation, to the social construction of reality. However, as Andrew Pickering (2010), Latour, and Bryant argue, the postmodern rejection of the two worlds actually reifies the subject-object dichotomy, privileging representation over substance, essentially reversing the modernist position, but not challenging its fundamental conceit. In other words, positioning language as that which constructs reality without admitting or engaging the agency of material forces allows the modernist position to remain intact.

Despite this rejection of representation and the potential for dissonance it creates, feminist scholars, especially those in STS, have significantly embraced NM.[4] This emerging body of scholarship finds common ground in the consensus that postmodern inquiry (even feminist postmodern inquiry) has too long privileged the discursive, eliding the material and the ontological. Indeed, preoccupation with the discursive and the attendant retreat from materiality is especially disconcerting to feminist scholars concerned with bodies, the environment, science, and socioeconomic realities. As Samantha Frost (2011) notes, NM offers feminist inquiry an opportunity to "re-introduce biological and material agency into feminist analysis" (p. 70). Similarly, Karen Barad (2003) argues, "What is needed is a robust account of the materialization of *all* bodies—'human' and 'nonhuman'—and the material-discursive practices by which their differential constitutions are marked" (p. 810). Here, the entanglement of forces, material and discursive, itself becomes the object of inquiry. Indeed, Stacy Alaimo and Susan Hekman's (2008) anthology aims to "bring the material, specifically the materiality of the human body and the natural world, into the forefront of feminist theory and practice" (p. 1).

However, this enthusiasm for NM within feminist STS is not without its tensions. As Alaimo and Hekman note, "materiality . . . has long been an extraordinarily volatile site for feminist theory" (p. 1). So too, Frost (2011) writes, "many [feminists] are likely to be suspicious of any 'biologizing' move that might, advertently or inadvertently, dress up power relations and disciplining norms as a force of nature or a biological imperative" (p. 71). Surely, biological essentialism has been incredibly damaging. For example, Manuela Rossini (2006) writes:

> When politicians and scientists have for centuries recurred to "natural" (because biological) differences to explain and legitimate social discrimination, oppression and inequality between the sexes and between human beings of different classes and ethnicities, it was more than necessary to counter, if not downright deny, biologistic argumentations.

We see in both Rossini and Frost that to merely admit, let alone inquire into, the agency of biological, physical, or chemical—the *material*—has been deemed by many as counterproductive to feminisms' rejection of biological essentialism.

While these concerns over the potential hegemonic uses of biology and the body as more than discursive are essential issues that must be addressed as a part of feminisms' engagement with NM, they are somewhat beyond the scope of this chapter. Rather, we aim to highlight another deeply problematic tension between NM and feminist commitments (in other words, the problem of political representation). As scholars ranging from Gayatri Chakravorty Spivak (1999) to Latour remind us, questions of political representation are intimately entangled with the problems of epistemic representation. Whether it is Spivak's exploration of the epistemic violence that underlies both scientific and political representation or Latour's analysis of Robert Boyle's air pump and Thomas Hobbes's *Leviathan* as avatars of the modernist project, there are highly compelling reasons to suggest that it would be very inappropriate to posit two entirely unrelated spheres of representation—one epistemic, one political. In particular, Latour's exploration of the reciprocal entanglements between the modernist political and scientific projects makes this unadvisable. Furthermore, disentangling these two spheres of representation replicates the modernist dualisms that separate science and the natural world from politics and the human.

Subsequently, it is incumbent on NM scholars with feminist commitments to tackle head-on the ramifications of NM's rejection of (epistemic) representation for feminist political projects. Assuredly, some variants better lend themselves to feminist rhetorical scholarship than others. And, it is precisely this issue we address in this chapter. In so doing, we begin with feminist new materialist Annemarie Mol and her argument that NM requires a transition from a (post-)modern politics of *who* to a politics of *what*. Using this argument as a foil, we then explore the affordances and limitations of politics of who versus politics of what inquiry for a long-standing area of feminist political concern: patient participation in health policy decision-making. Ultimately, in light of the complexities exposed by our comparative analysis, we argue that our adaptation of Mol, specifically multiple ontologies, offers a methodological approach that takes insights from NM, yet remains consistent with feminist political commitments.

Politics of Who versus Politics of What

In *The Body Multiple*, Mol (2002) argues that the ostensibly emancipatory efforts of postmodern inquiry into patient representation actually serve to disenfranchise patients and further marginalize the patient perspective in health care and health policy. In making this argument, Mol challenges us to transcend the disease/illness dichotomy, a long-standing staple of postmodern feminist inquiry. As she notes, the traditional postmodern approach posits two distinct entities:

> In addition to *disease*, the object of biomedicine, something else is of importance too, a patient's *illness*. Illness here stands for a patient's interpretation of his or her disease, the feelings that accompany it, the life events it turns into. (p. 9)

The problem here is that despite the disease/illness dichotomy's position as an invention of postmodern inquiry, it is a fundamentally modernist conceit. There is a singular biomedical reality about which patients have feelings. There is a disease about which people (doctors and patients) have different viewpoints. As Mol further elaborates,

> in the social sciences, "disease" and "illness" were separated out as two interlinked but separate phenomena. Social scientists put "illness" on the research agenda. Shelves and books and volumes of journals were dedicated to it. Interviews were amassed, the attribution of meaning was analyzed, and ways of therapeutically attending to it were designed. All along social

scientists left the study of disease "itself" to their colleagues, the physicians, until they started to worry about the power of a strong alliance with physical reality grants to doctors. Then, social scientists gradually began to stress that reality isn't responsible all by itself for what doctors say about it. "Disease" may be inside the body, but what is said about it isn't. Bodies only speak if and when they are made heavy with meaning. (pp. 9–10)

And, herein, lies the essential difference between the politics of who and the politics of what. Who/what should be made to speak? Who/what should be represented in political spaces? Under the postmodern perspectival project, people with perspectives must be represented, must be allowed to speak.

But why? Fundamentally, we understand the expressed need for patient representation in political/policy spaces as an effort to confront something very much like Spivak's epistemic violence in *A Critique of Postcolonial Reason* (1999). Biomedical researchers enlist scientific technologies from lab tests to patient outcomes inventories as a way of speaking for patients. You cannot speak for yourself, they say. Let us speak for you. Now obviously, this argument does not pass muster in feminist scholarship. It is the very model of scientific patriarchy in action. However, in their efforts to deconstruct the authority of patriarchal sciences, the postmodernists provided us with a "perspectival"[5] approach to the body. In so doing, as Mol notes, this approach commits its own act of epistemic violence. The body that cannot speak for itself must be represented by people with perspectives and made heavy with meaning. Following this logic, for example, people's perspectives on the MMR vaccine's link to autism take precedence over the dramatic increase in measles cases in the wake of anti-vaccination sentiment.

NM, on the other hand, requires a shift from this politics of who to a politics of what. As Mol (2002) writes,

a politics-of-what explores the differences, not between doctors and patients, but between various enactments of a particular disease.... [D]ifferent enactments of a disease entail different ontologies. They each *do* the body differently. But they also come with different ways of *doing* the good. In each variety of [a disease] the *dis* of the *dis-ease* is slightly different. Different, too, are the ideals that, standing in for the unreachable "health," orient treatment. (p. 176)

Here, Mol grounds the shift to a politics of what in her multiple ontological approach to atherosclerosis. Consistent with the new materialist favoring of ontology over epistemology, Mol's approach foregrounds the *practices* through which disease comes into being—enactments of disease(s). Mol uses *enact* to describe activities that "take place" and "practices" that stage multiple disease ontologies (p. 33). In Mol's case, for example, the disease enacted by a patient at home is different than the disease enacted by a pathologist and a microscope. She writes, "Since the object of manipulation tends to differ from one practice to another, reality multiplies. The body, the patient, the disease, the doctor, the technician, the technology: all of these are more than one" (p. 5). For Mol, this does not mean disease is plural (perspectives are plural); instead, what a pathologist does with a microscope and what a patient does at home manifests multiple diseases. Rather than many perspectives on a single biomedical reality, an ontological approach posits multiple diseases. Ultimately, this theoretical underpinning results, for Mol, in an ethnography of practices or praxiography. Of course, a praxiographic approach involves examining discursive practices, yet it is important to note that such practices do not constitute disease alone, but rather disease(s) is(are) enacted in material-discursive events.

Under this rubric, an ethical health policy forum must be engineered so as to include not only the testimony of people with perspectives but also (perhaps more importantly) the testimony of spaces, bodies, and practices. Complexities arise, no doubt, when the former contradicts the latter. And, of course, here is the strong potential tension with feminist political commitments. A requirement that embodiment is represented is not the same as a requirement that patients are represented. Nevertheless, an adoption of the NM approach and its attendant rejection of representation and the politics of who seems to require just such a move. This, of course, places NM scholars with feminist political commitments in a potentially difficult situation. The theoretical argument against representation is quite compelling, yet it directly challenges long-standing political commitments. And, ultimately, there is no easy escape from this theoretical-political morass. Reject NM and potentially be a backsliding (post-)modernist. Reject the politics of who and potentially perpetuate the marginalization of the disenfranchised.

Recognizing the lack of an easy answer here, we have opted instead for presenting a direct comparison of the politics of who and the politics of what. That is, in the sections that follow, we offer contrasting evaluations of the U.S. Food and Drug Administration's (FDA) patient representative (PR) program. This program serves as a primary vehicle for ensuring patient representation in health policy decision-making. As such, our contrasting analyses offer great potential in terms of helping to resolve these competing exigencies.

The FDA's Patient Representative Program

In the 1960s, the FDA established its first Drug Advisory Committee (DAC), and there are now seventeen different disease-domain specific DACs. These committees are expert advisory bodies designed to (*1*) provide independent evaluation of newly proposed pharmaceutical products and devices; (*2*) offer independent assessments of newly proposed uses for existing pharmaceutical products and devices; and (*3*) deliberate about and evaluate proposed methods for investigating pharmaceutical products and devices (U.S. Food and Drug Administration, "Human," 2014). DAC meetings are primarily forensic in nature. Two competing parties (a drug sponsor [the company that manufactures the drug] and an FDA team charged with playing devil's advocate) provide arguments and testimony and participate in cross-examinations before a committee of evaluators. Historically, DAC evaluators were exclusively medical experts (ME) hired to be independent analysts. However, since at least the 1980s, a great deal of scholarship in medicine, bioethics, and health policy has argued that patients and consumer stakeholders must be included in health policy deliberation and decision-making.[6] This scholarship echoes long-standing calls for stakeholder inclusion from a wide variety of patient and disease advocacy organizations.[7] A principal concern of these scholarly and advocacy efforts is to ensure that patients and consumers are adequately represented in policy forums traditionally dominated by biomedical researchers and health-care practitioners. In 1991, the FDA responded to such calls by integrating the PR program into its established DACs.

With the addition of the PR program, DACs constitute an important venue for experts—stakeholder engagement in pharmaceutical policy deliberation. According to the FDA, a primary role of PRs is to "provide [the] FDA with the unique perspective of patients and family members affected by a serious or life-threatening disease" ("About," 2014). The FDA's language here and its situation in the advocacy and advocacy scholarship of the 1990s clearly enact a postmodern politics of who. PRs are charged with providing the "unique perspective" of patients. They are charged with bringing illness into a deliberative domain typically dominated by disease. Now, the extent to which the FDA's management of the PR program actually enables PRs to bring this unique perspective is an outstanding question. Furthermore, it is equally unknown whether or not this politics of who approach is actually capable of implementing the underlying goals of ensuring that patient bodies and embodied practices are faithfully represented. Addressing these questions, of course, maps onto the primary goal of this chapter—to offer a comparative evaluation of the politics of who versus the politics of what. As such, the following sections first evaluate the success of the FDA's program on its own terms and then under the politics of what. That is, we first assess whether or not the PR program successfully manages to redress the imbalance between expert and patient stakeholders, and then we investigate the extent to which these efforts actually represent the whole range of disease practices. In so doing, we rely to a certain extent on quantitative analyses. We wish to highlight that such approaches are not without limitations, and that humanistic or critical/cultural approaches would capture something very different. Yet we feel the approach taken here is well suited to the program-wide evaluation for which we aim.

Further, we acknowledge that our approach to demonstrating the enactments of different disease ontologies (coding DAC meeting transcripts) may superficially seem at odds with the NM theory of disease enactment. However, as we argue in greater depth later, our focus on the spaces, bodies, and practices that emerge in the material-discursive event represents a viable new materialist rhetorical methodology.

The Politics of Who

In order to assess the extent to which PRs provide something of a balance against the dominance of MEs, we evaluated 167 of 187 DAC meetings held between 2009 and 2012.[8] For each meeting, we identified each PR and each ME on the DAC. Using these identifications, we then loaded each meeting transcript into the qualitative data analysis software NVivo for Teams. Using this software, we coded each DAC member according to their stakeholder status (ME or PR). This allowed us to easily tally the relative distribution of the stakeholder groups.

DAC meetings in the data set were found to feature between 8 and 38 total committee members. Member type data reported in the first table represent the distribution in relation to all committee members. In particular, we'd like to draw the reader's attention to the discrepancy between MEs and PRs. While MEs composed, on average, 81 percent of the committee members, PRs composed only 5.79 percent. Furthermore, on no occasion were more than two PRs present, and in 29 cases there were none. In total, there were 152 PRs and 2,172 MEs in the data set, or about 1 PR to every 14 MEs.

Table 6.1. DAC member type distribution

DAC member type	Minimum		Average		Maximum	
	No.	Percent	No.	Percent	No.	Percent
Patient representatives	0	0	1	5.79	2	15.79
Medical experts	6	67	14	81	36	97

Transcripts were further coded to identify every time each individual member talked during the course of each meeting. Both utterances and number of words were used to calculate the contributions of each member type relative to the others. In each meeting, one ME was excluded from the contribution analysis because in every meeting an ME is designated the meeting chair, giving him or her special responsibilities such as calling on other members to talk and reading the voting questions. While we excluded these data, for they would have unfairly raised the average contributions of MEs, it is important to note meeting chairs are responsible for granting committee members the opportunity to speak; a PR was never selected to chair.

As table 6.2 shows, individual PRs spoke significantly less than MEs with the average ME speaking almost double the words per meeting. The fact that there are on average 13 more MEs than PRs in the room becomes even more striking when one notes that individual MEs spoke more often and longer than PRs. While the FDA should be commended for implementing the PR program, our analysis shows (*1*) very few seats were allocated for PRs; and (*2*) PRs spoke significantly less than other committee members. Extending

Table 6.2. Individual stakeholder contributions per meeting

	Minimum	Average	Maximum
	Patient Representatives		
Utterances	1	10	57
Words	6	461	2,694
Percent	0	3	17
	Medical Experts		
Utterances	1	17	112
Words	1	928	12,645
Percent	0	5	42

notions of representation beyond the mere presence of a member of a stakeholder group to consider to what extent that constituency contributes calls into question the quality of the PR program as it currently functions. However, this is only a politics of who assessment.

Politics of What

Shifting our analysis from a politics of who to a politics of what requires we focus our attention not on who shows up and speaks but rather what spaces, bodies, or practices are enacted. The locus of analyses becomes practices—the doings or enactings that bring objects into and out of being. As Mol's praxiography demonstrates, disease is *done* or *enacted* differently in the exam room and in the lab, resulting in not different perspectives of a disease but *diseases*. In deploying Mol's approach, we follow the lead of prior work on a specific DAC meeting within a politics of what idiom. In their investigation of the public comment period of the 2011 Oncologic Drugs Advisory Committee hearing on the breast cancer drug Avastin, Christa Teston, S. Scott Graham, Raquel Baldwinson, Andria Li, and Jessamyn Swift (2014) make an important contribution to our understanding of DAC meetings. While they commend the FDA for including multiple stakeholders (for example, patients, allies, advocates), their analysis suggests that not all stakeholders were able to fully participate. They elaborate, suggesting that while the FDA recognizes that patients have a stake, it does not recognize their embodied expertise (p. 166). Indeed, identifying the lived experience of patients as a critical, yet marginalized, ontology in the Avastin hearing informs this work.

Additionally, one might infer within a politics of who frame that by either adding more seats for PRs or making more opportunities for them to participate, the problem of representation would be solved. However, drawing on Mol, Teston et al. (2014) identify four ontologies present in the Avastin public comment period (empirical-discursive, artful-practical, phenomenological, and capitalist). In identifying the relative prevalence of the various disease ontologies, the researchers demonstrate not only the marginalization of patient voices but also the marginalization of disease as lived/done/enacted outside the domain of evidence-based medicine. Compelling as it is, their analysis is confined to the public comment period of a single, atypical meeting. In extending this work, the current study represents a comprehensive analysis of four years of DAC meetings—an analysis that accounts for not only who is allotted for patient representation *but also what is enacted when they get there.*

Using Mol's sites of practice (2002), Teston et al.'s ontologies (2014), and S. Scott Graham's rhetorical-ontological inquiry (2015), we identified four primary ontologies in our analysis: lab, home, clinic, and market (defined in table 6.3 and described later in full detail). While looking at language in this way might initially seem like a traditional rhetorical analysis, we wish to make clear that we are chiefly interested in the practices from which ontologies emerge. To be sure, DAC meetings are primarily discursive events. However, the question for us does not center on speakers' linguistic strategies but rather the disease ontologies enacted in the discursive space. In other words, a politics of what requires interest not in who is speaking or how they speak but in *what* spaces, bodies, practices are made manifest. Culling insights from these authors, we have labeled our ontologies following the site of practice from which they primarily emerge.

Table 6.3. Sites of practice coding schema

Lab	Drug or disease* enacted in the lab through clinical trial data, statistical analysis
Home	Drug or disease enacted in the home, at the medicine cabinet
Clinic	Drug or disease enacted in the clinical exam room, patient-provider encounter
Market	Drug or disease enacted at insurance companies, pharmacies, the bank

*We use "drug or disease" to get at the complicated biological entanglement of pathology and pharmacokinetics. In clinical trials in which a drug to be used on a given medical condition is under adjudication, the home ontology often arises as experience with the "condition during treatment" is articulated. In such cases, the experience of the drug and the experience of the condition cannot be easily separated.

Lab. The disease arising from statistical assessments of clinical trial data enacts the lab ontology. Evidence-based medicine, clinical trials, pharmacokinetics, pathologies, hazard ratios, and deaths are the tools that bring this "disease" into being. As such, the lab ontology stages empirical statistical analysis as the appropriate practice through which health and pharmaceutical policy should be adjudicated. Indeed, the FDA is subject to Congress's 1962 amendments to the Food and Drug Act that require all federal pharmaceuticals marketed in the United States to "demonstrate safety and efficacy through 'adequate and well-controlled investigations'" (Graham, 2015, p. 58). In practice, then, evidence-based medicine, more specifically the randomized controlled trial, is the DACs' evidentiary gold standard. The lab is most often enacted as in the following statement:

> I'm still not sold on the fact of the efficacy for the EGFR IHC–negative and squamous cell. To me, it's weak and unproven. We already have drugs out there that are not beneficial to squamous cell at all. This one I'm not saying is not beneficial at all, just that I don't think it's proven that it's going to be beneficial. The efficacy just isn't there. (Moffitt, 2009)

Here, the lab ontology is enacted as "efficacy" and "benefit" are assessed in terms of clinical trial outcomes. Lab tests are conducted by investigators aimed to determine tumor growth, progression, regression, and so on in patients in response to a drug or therapy. These trials tend to involve many patients and investigators spread across a handful of the most prestigious medical centers in the United States and abroad. Further, such trials are both highly constrained in their scope and application and subjected to statistical analyses in service of the ideals of certainty and objectivity. In other words, they are designed such that patient-reported outcomes or doctor intuition are not factored into the measure of efficacy produced.

Home. The home ontology arises from the everyday lived experience with a disease, such as interacting with the prescription drug label or managing symptoms, side effects, or psychosocial impacts of disease/drugs. Generally, this ontology is staged through personal experiences as offered by the PR, though at times it is staged as the combined lived experiences of a particular patient population or group (sometimes by an advocate, family member, or doctor).[9] Following is one example of the home enacted:

> As a patient myself, I cannot imagine taking one dose and having to wake up two to four hours later to take another dose, simply because it's hard enough to get to sleep and it's hard enough once I am asleep to go back to sleep once I have awakened. (Berney, 2010)

This example enacts a drug/disease under adjudication much different than that of the lab ontology. Specifically, here the lived experience of the drug (in terms of dosing) becomes critical when assessing the potential value of the drug under adjudication.

Clinic. The clinic ontology appears from practices within the exam room and patient-provider encounter—primarily diagnosing, treating, and prescribing. Similar to the home, the clinic ontology presumes experiential evidence as not only relevant but also critical to assessing the potential value of the drug or treatment under adjudication. For example, the following statement regards a drug that requires "periodic" reevaluation by a physician to ensure its continued effectiveness and safety. The commenter argues:

> Unfortunately, that still means never. Basically, for most people, it'll never come up. I really believe you need to put some sort of—whether it's yearly—most people, if you go for a yearly physical and you think, okay, I need to—at this physical, I also need to talk to the doctor about bisphosphonate that I'm on or whatever, that that's an opportunity to do it. But this particular style of language is going to go in one ear and right out the other, if the person ever even reads it at all. (Tucker, 2011)

Presumably, lab data indicated a need for continued monitoring of patients using the given drug. However, no specific time period was dictated by the labeling (meant for patients and providers). Serving often as an intermediary between the lab and home ontologies, the clinic ontology arises from the site at which providers must use their acumen to calibrate clinical trial data to patient condition and home practices. In the aforementioned case, prescribing and monitoring practices combine with patient tendencies to enact the clinic as an important ontology for adjudicating the potential value of the given drug when used with the label as currently written.

Market. The market ontology emerges from practices that are more spatially distributed than the others. Whereas the other ontologies appear from predictably "medical" spaces, the market ontology arises from purchasing practices as well as concerns of pharmaceutical marketing and patents. To be sure, matters of affordability, accessibility, and the intricacies of insurance coverage are often a feature of DAC deliberation. Consider the following example:

> Prescribing of this drug, will that be covered by insurance? Is this drug going to be the same price as a drug without niacin? And I'm not taking exception to the objective of the company in formulating this product. I'm just wondering. It's only going to be helpful if you get it into the user's hands who needs it at the time they need it and they don't abuse it. Is an insurance company going to cover it the same as they do a prescription for regular oxycodone? Has that been looked into, the marketing aspects of it? (Dubbs, 2010)

In the passage, the market ontology emerges as the participant invokes the monetary considerations that must be accounted for in approving this drug. Certainly, drug approval is the first step in availability of a particular drug, but approval based on lab or home sites of practice does not ensure accessibility for patients. Therefore, such practices enact the market ontology as necessary to any drug/treatment approval.

Using this coding schema, all PR utterances were queried in NVivo resulting in a data set of 1,520 utterances. A sample (25 percent) of the PR utterances was first coded independently by two of the authors to ensure inter-rater reliability (table 6.4). Once reliability was confirmed, the remainder of the data set was divided and coded (table 6.5). Some utterances such as introductions and conversational niceties did not refer to any site of practice and were not assigned any code. Furthermore, given the complexity of DAC deliberations, multiple sites of practice were often discussed in a single utterance. As such, multiple codes were often attached to a given utterance. All totaled, 770 utterances received one or more codes. In reporting results, we first offer mono-ontological utterances and then multiple-ontological utterances.

Table 6.4. Inter-rater reliability

Ontology	Lab	Home	Clinic	Market
Kappa	.987	.947	.889	.894
Percent agreement	99.38	97.81	95.94	98.43

Table 6.5. Most frequently occurring ontologies by utterance

Ontology	Utterances (n)	Utterances (%)
Lab	422	54.8
Home and lab*	200	25.97
Home	163	21.17
Home and clinic	95	12.34
Clinic	80	10.39
Lab and clinic	78	10.13
Home, lab, and clinic	68	8.83
Market	21	2.72

*Colocates listed here occurred most frequently.

Particularly, we'd like to highlight the lab ontology as overwhelmingly the most present in PR mono-ontological talk. The fact that the lab ontology alone was enacted in more than half of the coded utterances may seem to suggest that PRs deem statistical analysis of randomized controlled trial data the appropriate ontology for making policy decisions. However, one must wonder to what extent structural, contextual, and material factors shape and constrain PR discourse. Indeed, the FDA in each meeting poses specific questions to the committee. The questions themselves typically enact the lab ontology. For example:

> Do the observed treatment differences—Humira 160/80/40 versus placebo—in the proportion of patients that had clinical remission at week 8 of 9.3 percent, with a 95 percent confidence interval of 0.8 percent to 17.9 percent in study 826, and 7.2 percent, with a 95 percent confidence interval of 1.3 percent to 13.2 percent in study 827, represent a clinically meaningful benefit? (Kumar, 2012)

To what extent does the ontology enacted here and throughout the meeting shape PR discourse? In what ways might PRs, who are ostensibly brought to the DAC meetings to provide "the unique patient perspective," move beyond the lab ontology to enact other sites of practice as relevant, appropriate, or important to patients? Here, we see the multiple-ontological codes as particularly salient.

Take for instance the high frequency of home in the dual and tri-coded utterances. Considering the broad dominance of the lab ontology in DAC deliberation, these codes seem to suggest attempts to enact the home as an appropriate site of practice in a context that elicits the lab. Take for example the following exchange between a PR and an ME (also, the chairperson):

> PR SHAYA: We have this question on the slide, has a favorable risk-benefit profile been demonstrated? But I think that actually hides the more pertinent question. . . . And that question is, is the risk-benefit profile favorable to me? And I don't know who that person is, I don't know anything about them, I don't know how old they are, do they have kids, what are their priorities, any of that. (Shaya, 2012)
>
> CHAIRPERSON WILSON: Thank you for those thoughts. One thing I would say—and I think I do understand what you are saying—the U.S. government of course legislates that the FDA has to approve drugs. And so at the end of the day, that approval hopefully is going to be based on the best evidence we have at the time that the risk-benefit is there. And so I think that's what everyone's grappling with right now. (Wilson, 2012)

While the PR here enacts the home ontology, staging an individual's lived experience of a disease as crucial to interpreting "risk" and "benefit," the chairperson enacts "risk-benefit" as the domain of lab ontology in what might be understood as a polite, subtle reprimand.

A politics of what analysis, then, suggests PRs enact lab ontology at higher rates than home, clinic, or market. While we cannot say with certainty, this seems to point to a systematic elision of non-lab ontologies. Indeed, PRs enact home without lab too, but the enactment of the home ontologies is more prevalent when paired with lab. Ultimately, this research would be productively extended through inquiry into the nature of ontological colocations. Such analysis, we believe, would provide further insight not only into conflicts between ontologies but also into how multiple conflicting ontologies might be productively enacted simultaneously, resulting in more effective and ethical DAC deliberation and decision-making.

The Politics of What and Cross-Ontological Calibration

So, ultimately, what do these two analyses tell us about the FDA, the PR program, and the politics of who and what more broadly? At the very least, our findings confirm Teston et al.'s suspicion that the FDA's enactment of patient representation is deeply problematic and falls short of the stated goals of the program. Furthermore, these findings also strongly suggest that the politics of who is inadequate on its own to ensure that the embodied experience of patients is represented in pharmaceutical policy deliberation. Indeed, our analyses have determined that PRs contribute significantly less than other DAC members when contribution is measured in terms of percentage of words and number of utterances. Furthermore, we have found that even when PRs do talk, they are not necessarily enacting the ontology we might expect. PRs overwhelmingly address the practices of the lab and much more infrequently discuss the home or even the clinic.

Despite these concerning findings, our politics of what analysis does contain something in the way of hope. For indeed, a postplural approach to health policy decision-making would encourage the kind of calibrating activities found in our collocate passages. Specifically, we understand many of these collocate passages to be moments of cross-ontological calibration. "Calibration" is Mol's term for practices that allow multiple ontologies to be staged alongside one another so the differences among them might be explored and adjusted. On cross-ontological calibration in the clinic, she writes:

> If test outcomes were listened to as if they were each speaking for themselves alone, they might get confined within different paradigms. The question whether different tests say the same thing or rather something

> different would not be answerable—indeed it could hardly be asked. The possibility to negotiate between clinical notes, pressure measurements numbers, duplex graphs, and angiographic images arises thanks to the correlation studies that actively make them comparable with one another. The threat of incommensurability is countered in practice by establishing common measures. Correlation studies allow for the possibility (never friction free) of *translations*. (2002, pp. 84–85)

This notion of cross-ontological calibration is taken up in more rhetorical ways by S. Scott Graham's *Politics of Pain Medicine* (2015). As he notes, for Mol, "calibration is a sort of meta-activity . . . [where] different forms of diagnostic practice [are] realigned, reinterpreted, rearticulated so that they may 'speak' to each other, so that translation may occur" (pp. 87–88).

While this sort of calibration is not directly addressed in Mol, Graham explores the potential for calibrating activities as they circulate through the different structures of FDA oversight. And while *The Politics of Pain* offers some analysis of the calibrating potential of DACs, it does not address the particularities of rhetoric in action we see here. A PR program built on a foundation of a politics of what might just be able to help ensure that the kind of calibrating discourse we see here is a much more significant feature of the DAC program.

Conclusion

Of course, any revisions to the DAC and PR programs must be implemented very carefully. Enacting an NM approach to representation in pharmaceutical policy certainly cannot be done without much consideration of the deep, complicated entanglements of the perspectival and the ontological, or rather, the who and the what. We would be very uncomfortable with a new PR program that directed PRs to speak only about those disease ontologies staged at home and not those staged in the lab. We would be even more uncomfortable with a DAC that made claims to adequately represent home- and clinic-based ontologies in the absence of a PR program. And we are wary of drug sponsors who leverage disease ontologies staged at home in the absence of strong efficacy lab data. Yet each of these disconcerting options is a plausible outcome of a shift from the politics of who to the politics of what. Thus, to fully disentangle and separate these patient perspectives from their "doings" or "enactments" is neither easy nor perhaps even entirely desirable.

Throughout the course of this study and the subsequent writing process, we, the authors, were continually struck by the difficulty that accompanies researching, understanding, and writing about representation within a NM idiom. We debated, struggled, and strongly disagreed about what PRs *should* be doing in these meetings, and, indeed, whether we are in a position to argue to make normative claims in this area. Our deeply held personal and political commitments to ensuring patient participation in pharmaceutical policy making were ever dissonant with a complete shift to a politics of what. Perhaps, in these moments, we failed to ultimately overcome the legacy of postmodern inquiry as NM inquiry suggests we should.

Yet in offering both a who and a what analysis, it becomes clear that though conflicts may arise, such approaches may be productively integrated. Indeed, a health policy forum assessed only in terms of a politics of who risks neglecting the importance of integrating the full range of practices relevant to pharmaceutical adjudication. At the same time, a health policy forum assessed only in terms of a politics of what risks erasing the already marginalized voices of patients. While some new materialists would no doubt reject a politics of who entirely, feminist science studies scholars might find a synthesis between who and what a politically productive tool.

Examining and writing about DAC meetings, or any deliberative forum, within a NM idiom presents unique challenges. Indeed, a preoccupation with the discursive is consistently a primary concern. While some rhetoricians will undoubtedly find this critique hostile and unwarranted, we agree that it is dangerous to become preoccupied with the discursive at the neglect of the material, or perhaps more accurately, we find that the two cannot and should not be uncritically disentangled. As such, rhetoricians, and more specifically feminist rhetoricians of science, are uniquely qualified to engage material-discursive spaces such as health and science policy forums.

Ultimately, despite these difficult questions and concerns, we feel we offer this project as a potential solution or alternative to the opposition of incommensurable perspectives in pharmaceutical policy deliberation. In so doing, our ultimate goal here is to provoke a much-needed conversation about the affordances and detriments of NM for feminist science studies and feminist rhetorics of science. To embrace NM inquiry without caution, it seems, is to reject long-held political commitments. These are difficult challenges, challenges that warrant significant attention. To that end, we have tried to bring NM inquiry and feminist political commitments into dialogue with one another so as to highlight some of the detriments and affordances of each.

Notes

1. See Bennett, 2010; Bogost, 2012; Bryant, 2011; Coole and Frost, 2010; Graham, 2015; Harman, 2009; Pickering, 2010; Rickert, 2013.

2. Notable examples include Bryant, 2011; Graham, 2015; Mol, 2002; Pickering, 2010; Rickert, 2013.

3. See Bryant, 2011; Graham, 2015; Harman, 2009; Pickering, 2010.

4. See Alaimo and Hekman, 2008; Barad, 2003; Frost, 2011; Grosz, 2010; Rossini, 2006.

5. Following Mol's use of the term, which we elaborate on later, we use "perspectival" here and throughout the chapter to refer to the way the politics of who, even its postmodern instantiations, authorizes a modernist theory of looking at/having a perspective on a singular, objective reality.

6. See Alderman, Hipgrave, and Jimenez-Soto, 2013; Hunink et al., 2001; Lewis, 2000; Lewis, Shadish, and Lurigio, 1989; Macpherson, 2004, 2006; Milewa, 2008.

7. For example, Bastian, 1998; Burton, 2005; Earp, French, and Gilkey, 2008; Tomes, 2006; Traulsen and Almarsdóttir, 2005; Wilkinson, 2008.

8. Twenty transcripts were excluded for reasons ranging from transcription formatting errors to atypical meeting formats; approximately 89 percent of available transcripts were coded.

9. Whether people who have not experienced the medical condition to be treated could enact the home ontology was a point of lively discussion throughout the study. In the end, we determined enactment of the home according to the specificity of the practices discussed. If specific practices were articulated, the home ontology was considered in isolation of the speakers' status as a patient.

References

Alaimo, S., & Hekman, S. (Eds.). (2008). *Material feminisms*. Bloomington: Indiana University Press.

Alderman, K. B., Hipgrave D., & Jimenez-Soto, E. (2013). Public engagement in health priority setting in low- and middle-income countries: Current trends and considerations for policy. *PLoS Med, 10*(8). Accessed 15 March 2014. Retrieved from http://www.ncbi.nlm.nih.gov/pmc/articles/PMC3735456/

Barad, K. (2003). Posthumanist performativity: Toward an understanding of how matter comes to matter. *Signs, 28*(3), 801–831.

Bastian, H. (1998). Speaking up for ourselves: The evolution of consumer advocacy in health care. *International Journal of Technology Assessment in Health Care, 14*(01), 3–23.

Bennett, J. (2010). *Vibrant matter: A political ecology of things.* Durham, NC: Duke University Press.

Berney, B. (2010, August 20). Comment made during public meeting of the Arthritis Advisory Committee. Bethesda, MD.

Bogost, I. (2012). *Alien phenomenology, or what it's like to be a thing.* Minneapolis: University of Minnesota Press.

Bryant, L. R. (2011). *The democracy of objects.* Ann Arbor, MI: Open Humanities Press.

Burton, B. (2005). Drug companies told that sponsoring patients' groups might help win approval for their products. *BMJ, 331*(7529), 1359.

Coole, D., & Frost, S. (2010). Introducing the new materialisms. In D. Coole & S. Frost (Eds.), *New materialisms: Ontology, agency, and politics* (pp. 1–43). Durham, NC: Duke University Press.

Dubbs, R. (2010, April 22). Comment made during joint public meeting of the Anesthetic Life Support Drugs Advisory Committee and the Drug Safety and Risk Management Advisory Committee. Gaithersburg, MD.

Earp, J. A. L., French, E. A., & Gilkey, M. B. (2008). *Patient advocacy for health care quality: Strategies for achieving patient-centered care.* Burlington, MA: Jones and Bartlett Learning.

Frost, S. (2011). The implications of the new materialisms for feminist epistemology. In H. E. Grasswick (Ed.), *Feminist epistemology and philosophy of science: Power in knowledge* (pp. 69–83). New York, NY: Springer.

Graham, S. S. (2015). *The politics of pain medicine: A rhetorical-ontological inquiry.* Chicago, IL: University of Chicago Press.

Grosz, E. (2010). Feminism, materialism, and freedom. In D. Coole & S. Frost (Eds.), *New materialisms: Ontology, agency, and politics* (pp. 139–157). Durham, NC: Duke University Press.

Harman, G. (2009). *Prince of networks: Bruno Latour and metaphysics.* Melbourne, Australia: re.press.

Hekman, S. (2008). Constructing the ballast: an ontology for feminism. In S. Alaimo & S. Hekman, (Eds.), *Material feminisms* (pp. 85–119). Bloomington: Indiana University Press.

Hunink, M. G. M., et al. (2001). *Decision making in health and medicine: Integrating Evidence and Values* (Vol. 1). Cambridge, UK: Cambridge University Press.

Kumar, A. (2012, August 28). Comment made during public meeting of the Gastrointestinal Drugs Advisory Committee. Silver Spring, MD.

Latour, B. (1993). *We have never been modern.* (Catherine Porter, Trans.). Cambridge, MA: Harvard University Press.

Lewis, C. (2000). Advisory committees: FDA's primary stakeholders have a say. *FDA Consumer, 34*(5), 30–34.

Lewis, D. A., Shadish, W. R., & Lurigio, A. J. (1989). Policies of inclusion and the mentally ill: Long-term care in a new environment. *Journal of Social Issues, 45*(3), 173–186.

Macpherson, C. C. (2004). To strengthen consensus, consult the stakeholders. *Bioethics, 18*(3), 283–292.

Macpherson, C. C. (2006). Healthcare development requires stakeholder consultation: Palliative care in the Caribbean. *Cambridge Quarterly of Healthcare Ethics, 15*(3), 248–255.

Milewa, T. (2008). Representation and legitimacy in health policy formulation at a national level: Perspectives from a study of health technology eligibility procedures in the United Kingdom. *Health Policy, 85*(3), 356–362.

Moffitt, P. (2009, December 16). Comment made during public meeting of the Oncologic Drugs Advisory Committee. Gaithersburg, MD.

Mol, A. (2002). *The body multiple: Ontology in medical practice.* Durham, NC: Duke University Press.

Pickering, A. (2010). *The cybernetic brain: Sketches of another future.* Chicago, IL: University of Chicago Press.

Rickert, T. (2013). *Ambient Rhetoric: The attunements of rhetorical being.* Pittsburgh, PA: University of Pittsburgh Press.

Rossini, M. (2006). To the dogs: Companion speciesism and the new feminist materialism. *Kritikos, 3.* Accessed 15 March 2014. Retrieved from http://intertheory.org/rossini

Shaya, S. D. F. (2012, March 20). Comment made during public meeting of the Oncologic Drugs Advisory Committee. Silver Spring, MD.

Spivak, G. C. (1999). *A critique of postcolonial reason: Toward a history of the vanishing present.* Cambridge, MA: Harvard University Press.

Teston, C. B., Graham, S. S., Baldwinson, R., Li, A., & Swift, J. (2014). Public voices in pharmaceutical deliberations: negotiating "clinical benefit" in the FDA's Avastin hearing. *Journal of Medical Humanities, 35*(2), 149–170.

Tomes, N. (2006). The patient as a policy factor: A historical case study of the consumer/survivor movement in mental health. *Health Affairs, 25*(3), 720–729.

Traulsen, J. M., & Almarsdóttir, A. B. (2005). Pharmaceutical policy and the lay public. *Pharmacy World and Science, 27*(4), 273–277.

Tucker, E. (2011, September 9). Comment made during joint public meeting of the Reproductive Health Drugs Advisory Committee and the Drug Safety and Risk Management Advisory Committee. Adelphi, MD.

U.S. Food and Drug Administration. (2014). About the FDA patient representative program. *FDA.* Accessed 15 March 2014. Retrieved from www.fda.gov/ForPatients/About/ucm412709.htm

U.S. Food and Drug Administration. (2014). Human Drug Advisory Committees. *FDA.* Accessed 15 March 2014. Retrieved from www.fda.gov/AdvisoryCommittees/CommitteesMeetingMaterials/Drugs/default.htm

Wilkinson, E. (2008). Patient organisations aim for greater collaboration. *Molecular Oncology, 2*(3), 200–202.

Wilson, W. (2012, March 20). Comment made during public meeting of the Oncologic Drugs Advisory Committee. Silver Spring, MD.

7.
Embodied Vernacularity at the FDA

Feminism, Epistemic Authority, and Biomedical Activism

Liz Barr

Despite current HIV prevention efforts, approximately forty thousand people become HIV positive every year in the United States. Although the HIV incidence rate declined nearly 20 percent from 2005 to 2014 (Centers for Disease Control and Prevention, 2017) and current HIV incidence is dramatically lower than rates of infection in the 1980s and 1990s, we remain frustratingly unable to eliminate new infections. As part of efforts to reduce these forty thousand new infections, many HIV prevention agencies in the United States are devoting more funds to prevention research with "high-risk" populations. Prevention is also shifting toward a harm-reduction model, key components of which are encouragement of condom use, HIV testing, treatment as prevention, and needle exchange programs. However, prevention experts are desperate for new strategies to reduce infection rates.

One promising prevention strategy is pre-exposure prophylaxis, or PrEP. The premise behind PrEP is that an HIV-negative person takes an antiretroviral drug daily. Over time, the drug builds up in the person's system and can have a protective effect against HIV infection.[1] HIV prevention experts and activists advocated for PrEP (or something like it) for several reasons. First, if existing prevention methods and models are falling short, PrEP might do something to reduce the forty thousand new infections each year. Second, PrEP can be used discreetly, without one's sexual partners' knowledge. Third, clinical research has yet to identify an effective vaccine against HIV and AIDS, and so PrEP is seen as an excellent prevention option until a vaccine is found. Concerns about PrEP generally focus on four areas: the ethics of making expensive HIV drugs available to HIV-negative persons when not every HIV-positive person has access to them; the potential for resistant strains of HIV to develop if someone taking PrEP becomes HIV positive; long-term toxicity; and the willingness of "healthy" persons to consistently take a drug with sometimes severe side effects.[2] Advocates and critics of PrEP came together on May 10, 2012, for a U.S. Food and Drug Administration (FDA) hearing on the pharmaceutical company Gilead Sciences' "new drug use" application to market one of its existing HIV drugs—Truvada—as PrEP.[3] On July 16 of that year, the FDA announced that Truvada had been approved for this new use.

Because of earlier community-based activism by groups such as the AIDS Coalition to Unleash Power (ACT UP), the federal government mandates that the FDA seek community input at hearings like the one described as well as in the clinical research protocol design process (Kagan et al., 2012). The Truvada hearing was no exception. Although this mandate was followed—space was set aside for community testimony, there were community members in the audience, and three community representatives sat on the twenty-two-member committee—the hearing was an institutional event, physically and discursively located within the boundaries of the FDA. An overwhelming majority of the voices that spoke at the hearing were those of clinical researchers and pharmaceutical company representatives. These voices arrived at the table with a sense of legitimacy and rhetorical power by virtue of their institutional affiliations and the cultural trust placed in scientific expertise.[4] Because of the inclusion of community testimony, however, the hearing contained elements of the vernacular, or noninstitutional, as well. The community members had some claim to legitimacy and authority because of their invited presence at the hearing, but their position as nonexperts meant that they began from a position of *less* authority than the scientific voices. As such, the community voices embodied a disempowered position that required them to employ different rhetorical strategies in order to be heard.

As we think in this volume about the intersections between rhetoric and feminist science studies, the Truvada hearing emerges as an excellent case study of the rhetorical strategies vernacular actors employ to claim authority in asymmetrical relationships in the scientific sphere. Grounded in this case study, this chapter explores a rhetorical concept I term "embodied vernacularity," which accounts for the speaking body in addition to the spoken word. I do not use this chapter to argue a particular position on PrEP (although I do think it is an important HIV prevention strategy and am happy to see it gaining traction). Instead, my goal here is to draw on the case study in order to demonstrate how embodied vernacularity can contribute to research in feminist rhetorical science studies. To do so, I focus on the strategies used by community rhetors at the Truvada hearing. I argue that community members developed an embodied vernacular authority that positioned itself as an alternative to the institutional, scientific discourses that dominated that day. This embodied vernacularity drew on a different set of material and symbolic resources (including individual and specific bodies moving in the world and bodies having sex) than those used by the dominant discourses (bodies aggregated as data points). Whether they were pro- or anti-Truvada, many vernacular rhetors used their bodies (and the bodies of fellow community members) as argumentative resources in an attempt to influence the committee's policy decisions.

Of course, the community rhetors were not alone in their use of embodied discourse. Scientists at the hearing also spoke from within fleshy bodies, and they too brought bodies into their arguments. However, unlike the scientists who spoke of a bloodless body composed of aggregated information, many community speakers referenced a visceral body, establishing their ethos through testimony that accounted for what was physically happening to their bodies due to Truvada. By attempting to reframe the terms of the Truvada debate through the body's knowledge, the community testimony enacted a feminist and rhetorical challenge to the norms of scientific discourse. Feminist science studies foregrounds the materiality and relationality of the bodies involved in knowledge production (Clough, 2013; Willey, 2016). Rhetoricians working on science and medicine have demonstrated how scientific communities undertake complex rhetorical maneuvering to maintain the cultural power of scientific authority (Condit, 1996; Reeves, 1992, 1998), and the community testimony at the Truvada hearing reveals one strategy—embodied vernacularity—through which vernacular actors might challenge this dominance.

Embodied vernacularity (which I situate within rhetoric scholarship in the next section) has roots in feminist theories, including its awareness of power asymmetries and attention to strategies for resistance (hooks, 2000; Novotny, 2015); rejection of the neat bifurcation between mind and body (Ahmed, 2006; Clughen, 2014); and recognition that bodies—both their materiality and their movements through the world—are valuable sources of knowledge (Clare, 2003; Moraga & Anzaldúa, 1983). Analysis of the rhetorical strategies community rhetors employed in their statements reveals that many of these speakers saw their bodies not as instruments of biomedical data production but as valuable sites of authority in their very fleshiness. Whether countering aggregated biomedical data with personal narratives or foregrounding the complications and contradictions that can accompany having sex with a serodiscordant[5] partner, vernacular rhetors enacted Cherríe Moraga and Gloria Anzaldúa's "theory of the flesh." As Moraga and Anzaldúa (1983) explain, their theory of the flesh posits that "the physical realities of our lives . . . fuse to create a politic born out of necessity." (p. 23). A theory of the flesh thus centers material bodies as they both navigate the power inequalities that govern our world and seek social and cultural change. Likewise, embodied vernacularity focuses on fleshy bodies claiming space to counter dominant institutional discourses. If feminist theories of the body facilitate explorations of power and resistance, rhetorical theories of embodiment enable analyses of the recursive relationship between discourse and materiality and the mechanisms by which particular bodies are able (or unable) to make arguments or draw on inventional resources.[6] A feminist rhetorical perspective like embodied vernacularity, then, is attentive to the realities and the potentials that bodies encounter as they move within and against institutions.

Embodied vernacularity also requires acknowledgment of the specific location from which one speaks and the particular body that is doing the speaking. Donna Haraway (1988) tells us that "only those occupying the positions of the dominators are self-identical, unmarked, disembodied, unmediated" (p. 586). The ability to remain "unmarked" is a position of privilege; ignoring embodied differences reinforces systems of domination. Haraway's situated knowledge, an alternative feminist positioning, instead reveals webs of connection between the particular and the global. From this positioning, "feminist embodiment . . . is not about fixed location in a reified body, female or otherwise, but about nodes in fields, inflections in orientations, and responsibility for difference in material-semiotic fields of meaning" (p. 588). Embodied vernacularity is fundamentally a form of situated knowledge. Further, while gender is not the center of this chapter's analysis, the methods and analysis are inherently feminist: attentive to power asymmetries and seeking possibilities for resistance.

In what follows, I discuss theories of the vernacular and rhetorics of the body, elucidating the ways these two theoretical frameworks function in concert to create embodied vernacularity. Then, I turn to the power asymmetries present at the Truvada hearing, exploring the efficacy of embodied vernacularity as part of a hybrid strategy of resistance. Through examination of two major themes in community members' testimony—bodies in the world and bodies as sexual objects and subjects—I am able to suggest what embodied vernacularity offers us, even when it "fails."

Vernacular Discourse and Embodied Rhetoric

The vernacular is generally understood to be synonymous with everyday, noninstitutional verbal communication. For example, Gerard Hauser (1999) uses *vernacular* to refer to common discourse available to all. This type of vernacularity includes events such as quotidian conversations on street corners or conversations about politics between members of the public. As Kent Ono and John Sloop (1995) use the term, the vernacular is discourse produced by subaltern and disempowered groups. This type of vernacularity is defined by the positionality of who is speaking. Robert Glenn Howard (2010) expands vernacularity to a dialectic process that "emerges . . . whenever expressive discourse situates itself as alternate to the institutional" (p. 249). For Howard, vernacular discourse carries traces of the institutional, as the existence of the former depends on its relationship to the latter. There is, for Howard, no purely vernacular discourse—it always already exists in relation to institutional discourse. Further, unlike Hauser, who locates vernacularity in the content of discourse, and Ono and Sloop, who locate it in the identity of the rhetor, Howard argues that it is the *site and content* of discourse that makes it vernacular. In other words, a rhetor must assert an alternate, oppositional authority that is inherently related to the authority of institutional discourse. This framework is especially salient for my analysis of the FDA advisory committee meeting. The community rhetors' testimony—which countered the institutional arguments of the scientists—emerged as "alternate to the institutional," yet it could do so only by virtue of its relationship to the authority asserted by institutional rhetors.[7]

Howard's conception of vernacularity is valuable in that it elucidates the nuances of the vernacular-institutional relationship. However, rhetoricians can further expand our understanding of the vernacular beyond the discursive to account for the materiality of bodies. Embodied vernacular theory focuses on expressive *bodies* rather than expressed discourse. As Kevin DeLuca (1999) argues, bodies are not simply vessels for transmitting discourse; rather, they can persuade through their very physicality. Like many other rhetoricians, I am unconvinced that we can separate our words/minds from our bodies and argue for an embodied vernacular that acknowledges this tension (Dolmage, 2009; McKerrow, 1998; Price, 2015). In so doing, I align myself with rhetoric scholars who have argued against rhetoric's tradition of devaluing, ignoring, and erasing bodies by instead studying rhetoric as an embodied phenomenon (Chávez, 2009; Dolmage, 2009; Jordan, 2004; Shome, 2003). As Raymie McKerrow (1998) explains, "a corporeal rhetoric forsakes oppositionality in favor of an all-encompassing perspective on the rhetorical act" (p. 319). That is to say, focusing on the corporeal aspect of rhetoric allows us to dismantle the mind-body split and more fully understand a wide variety of rhetorical actions. Further, by accounting for materiality—embodiment and physical presence—we become able to explore more deeply relationships of power. Corporeal rhetoric resonates with feminist theories of embodiment, particularly when power, domination, and resistance are used as lenses of rhetorical analysis. Arguments are made by and through bodies that are located in systems of power, and our corporeality is influenced by the spaces we inhabit (Butler, 1993; Shome, 2003). Paying attention to arguing *bodies* results in a more complex picture of the argumentative terrain within which these bodies are embedded.

Understanding this argumentative terrain is significant for what J. Blake Scott might call an "ethics of responsiveness." Writing about HIV testing, Scott (2003) defines responsiveness as being both "accountable and caring: accountable to the needs, concerns, and contexts of those most affected by [HIV] testing and caring about meeting these needs and alleviating these people's harm and suffering" (p. 232). A feminist rhetorical approach to the Truvada hearing similarly suggests that an HIV prevention strategy should care about and be accountable to the people it is intended to protect. While scientific data is crucial to making decisions about biomedical technologies, so too are data that come from community members' lived experiences. And these experiences are embodied. Paying attention to bodies in the world, bodies as individuals, bodies as concrete and fleshy can illuminate spaces for doing science differently. We ought not take an either/or approach, throwing out biomedical data or erasing community expertise. Reading the Truvda hearing through the body reveals a both/and framework that hybridizes rhetorical strategies from both camps to create new epistemologies rooted in the lived experiences of bodies moving through the world.

Institutional Voices and Embodied Vernacularity

The FDA's Antiviral Drugs Advisory Committee met Thursday, May 10, 2012, from 8:00 A.M. to 8:30 P.M. The committee was composed of twenty-two members, some of whom were temporary members added on specifically for this hearing. Of these twenty-two members, three represented the community and the rest were MDs, PhDs, RNs, and other institutionally affiliated actors. The advisory committee was convened to hear testimony about Truvada's efficacy as PrEP and to make a decision whether to approve the drug for this new use. The bulk of the meeting (about ten and a half hours) was devoted to scientific presentations from HIV researchers from the Centers for Disease Control and Prevention, the National Institutes of Health, and Gilead Sciences. These presentations used the type of scientific language that one would expect to hear at a scientific conference, and they clearly presumed everyone present that day shared the same vocabulary.

Consider, for example, how Robert Grant, the principal investigator in one of the key studies being used to justify Truvada's efficacy as PrEP, explained his data:

> This is the plot of cumulative HIV infection over weeks since randomization. . . . P-value for nonproportional hazards was .43. . . .
>
> This is an analysis of efficacy by subgroup, defined by baseline characteristics. You can see there's trends toward efficacy in subgroups defined by age, level of education, region, Andean versus non-Andean, and alcohol use. (U.S. Food and Drug Administration, 2012, p. 100)

While there is nothing factually wrong in Grant's presentation of these data, the language he used reflects the erasure of individual characteristics—and thus disembodiment—common in biomedical discourse, which speaks of general rather than specific bodies. The lives of trial participants are reduced to demographic subgroups, P-values, and graphs. His emphasis on the data erased the people from whom the data were collected. Grant may as well have been talking about an experiment done in a test tube. While Grant's approach makes sense, given his positionality as a scientific rhetor, acknowledging the complex social worlds that trial participants inhabit would have allowed him to account for the social factors that affect data as well—busy schedules that may affect adherence to a daily pill regimen, shame or stigma about having a bottle of Truvada in one's medicine cabinet, or reluctance to experience the sometimes severe digestive side effects of Truvada. Of course, Grant needed to convey the data to his audience; however, his language choices reflect who he believes is his most important audience—the scientists and other institutional agents on the advisory committee and in attendance, not the community members also present in both capacities. These data were not presented in language that would be easy for lay community representatives on the committee or in the audience to understand. While many lay community members possess a great deal of scientific literacy, Grant required his audience to do all the work necessary to translate his scientific/institutional discourse into lay/vernacular discourse. Further, he privileged scientific fact over lived experience.

Institutional voices like Grant's dominated the Truvada hearing and drew heavily on biomedical and scientific authority. In traditional biomedical discourses, disembodied and depersonalized bodies provide evidentiary support for scientific hypotheses, as seen in the testimony earlier. Bodies are useful to these discourses only to the extent that they can provide valuable data; bodies therefore lack subject status as they are viewed only as objects for data collection. Most biomedical discourses develop ethos through traditional means, drawing on dominant culture's tendency to imbue medical and scientific professionals with blanket authority on medical and scientific matters such that community rhetors' authority and expertise are overlooked (Epstein, 1995). The majority of the testimony at the advisory committee meeting reflected this sort of authority: it ignored bodies' deep enmeshment in social, cultural, and relational arenas. Further, the scientists' privileged location excused them from having to account for their own embodiment (and embodiment's impact on knowledge production), allowing them to focus instead on the objective data gained from others' aggregated data points. Bolstered by the traditional biomedical authority on which the majority of the speakers relied, scientific fact reigned supreme.

Although community members were not included in the formal presentations, they were provided with the opportunity to speak for three minutes at a time during two hours in the afternoon. Community members were strictly held to this time limit (while scientific presentations often ran over their allotted time).[8] During these three-minute testimonials, people with HIV, their physicians and nurses, workers from the nonprofit sector, and ministers shared their opinions regarding why Truvada should—or, more often, should not—be approved as PrEP.[9] Unlike the scientific presentations, these community testimonies frequently centered embodied knowledge. For these rhetors, the mere presence of their bodies as well as their expressed bodily experiences and knowledges were sources of authority. As such, these community testimonies reflected a different understanding of what it means to have HIV, be at risk for HIV, and prevent HIV.

Power Asymmetries and Expertise

The embodied strategies that vernacular rhetors employed shifted the discourse of the day—for two hours at least—from objective science to lived experience. However, the power asymmetries between scientific knowledge and community expertise proved difficult to overcome. Community speakers struggled to establish their ethos in a rhetorical situation dominated by institutional, scientific discourse through their strategies of sharing personal experience, expressing real-world concerns about adherence, and questioning the potential implications of Truvada's side effects. Vernacular rhetors also struggled to counterbalance the weight of the dominant institutional discourse that framed people at risk for HIV as vulnerable and in need of protection from omniscient and omnipotent scientists and physicians, sometimes employing hybrid rhetorical strategies to do so. This hybrid approach relied on vernacular and institutional authority to stake claims. Rhetors choosing this strategy drew on both institutional discourse and vernacular expertise as inventional resources. While the institutional discourses they used unsurprisingly erased bodies, this erasure was mitigated by their embodied vernacular testimony.

This strategy is exemplified in the words of "Michael McConnell,"[10] who delivered powerful testimony based on his experiences living through the HIV epidemic. McConnell's use of a hybrid strategy was certainly shaped by the structure of the hearing. At the start of the community testimony, Judith Feinberg, the committee's chairperson, had reminded speakers of their time limit and cautioned them against "disruptive behavior such as shouting out" (U.S. FDA, 2012, p. 252). McConnell was the eighteenth community speaker,[11] so he had watched Dr. Feinberg strictly enforce the three-minute time limit for community speakers. To navigate the power asymmetries structuring the meeting, McConnell used his ethos as a gay man, someone who lost a lover to AIDS-related complications, and someone who is HIV positive to develop a vernacular authority that responded to the pro-PrEP institutional discourse:

> I want to thank the members of the committee for holding this hearing and allowing my VOICE [emphasis in original transcript] to be included in the evolving debate surrounding this important treatment as prevention issue.
>
> I have witnessed the full progression of HIV since its inception, including losing a partner to AIDS on a cold San Francisco morning in 1988. During the ensuing 24 years of living with HIV, I've taken almost every cocktail of meds imaginable as they became available until my virus eventually mutated a resistance to them, including Truvada. . . .
>
> . . . [M]y 12-year personal experience with real-life challenges that make adherence to taking HIV meds almost impossible, even when those whose very lives depend on it, are significant facts that render the current PrEP proposal useless in protecting anyone against existing Truvada-resistant HIV. More importantly, they legitimize concerns of whether or not the 44 percent efficacy achieved during the original PrEP trials, which were operated with controls unlikely to occur in real-life applications, are far more optimistic than attainable.

Therefore, I implore the FDA to consider the adherence experienced voice of reason for denying approval of PrEP until more research justifies the risks and costs to public health policy, especially at a time when limited existing funds should be more wisely spent on treating those already infected who are languishing on the nation's shameful ADAP [AIDS Drug Assistance Program] waiting lists. (pp. 289–290)

McConnell built ethos through his body's physicality; he was one of the few community members who raised his voice during his testimony (reflected in the transcript with capitalized words). By referencing the drug-resistant mutations of HIV that he has developed over the last twenty-four years, McConnell also used his own body as a source of data. However, unlike the disembodied scientific data that was presented at other points during the hearing, McConnell's use of body-as-evidence was rooted in his "personal experience with real-life challenges." McConnell constructed an embodied, experience-based ethos that relied both on fluency in the scientific data and emotional appeals only made possible by McConnell's status as a long-term HIV survivor.

McConnell also located the PrEP debate in a larger social context, reminding the committee that science does not take place in a vacuum and bodies exist in multiple realms. He noted the "limited existing funds" and the nation's "shameful ADAP waiting lists," reinserting the community's bodies into the discussion. We must remember that while PrEP is fundamentally about preventing HIV transmission, this prevention happens in the context of communities and bodies—bodies engaged in sexual activity and otherwise moving through the world. McConnell raised the ethics of protecting some bodies at the expense of others.[12] This sort of disparate distribution of Truvada might be seen as part of the long history of devaluing the bodies of people with HIV and rendering them expendable (Crimp, 1987; Treichler, 1999). Further, McConnell noted that continued government inaction, as evidenced by insufficient funds for treatment, contributes to the perpetuation of HIV infections. McConnell's testimony forces us to remember that bodies must be considered in context, including the context of government policies, when implementing new HIV prevention methods.

Perhaps most importantly, McConnell's testimony engaged in an interesting oscillation (Brouwer, 2001) between establishing authority from his lived experiences (he has "witnessed the full progression of HIV") and through his familiarity with scientific data (questioning "the 44 percent efficacy achieved during the original PrEP trials, which were operated with controls unlikely to occur in real-life applications"). He was clearly a vernacular actor, as reflected in his language and status as nonscientist. However, drawing on scientific knowledge from an embodied, vernacular position allowed him to make richer knowledge claims. This approach embodies Howard's conception of the vernacular as a hybrid position: it is both different and inseparable from the institutional. Employing this hybrid strategy, McConnell used two powerful yet distinct sources of knowledge—lay expertise and scientific expertise (Epstein, 1995). But most strikingly, McConnell moved beyond a hybrid *position* and rooted both strategies in his body and the bodies of other HIV-positive people. From his partial perspective and through his embodied vernacular authority, McConnell called the scientists' predictions about adherence into question.

McConnell's hybrid strategies resonate with Howard's understanding of the vernacular as something that emerges in opposition to the institutional yet remains tethered to it. But McConnell is also creating a theory of the flesh: using his own situated knowledge to claim authority and credibility. Theory of the flesh develops as we reckon with the messiness of our lives (for example, the "real-life challenges" that prevent optimal drug adherence). But theory of the flesh does more than catalog these individual contradictions; Moraga and Anzaldúa (1983) insist that theory of the flesh is part of a process of *social healing* (p. 23). McConnell makes links between his lived experiences and social policy, calling for the committee to recognize that his body's partial perspective is key to its authority.

Further, McConnell's testimony embodies Donna Haraway's (1988) "view from somewhere," or the idea that through multiple situated, partial perspectives we can find "connections and unexpected openings" (p. 590). McConnell's testimony offers a means of resisting the dominance of institutionalized science. He implicitly challenged the validity—and ethics—of extrapolating clinical trial results to populations *without* factoring in the complexities of individual people's lives. By making these connections through and with his body (his "adherence experienced *voice* of reason"), McConnell leveraged the power asymmetries present in the hearing to challenge what counts as expertise. Although McConnell did not articulate his position as a feminist one, reading his testimony through feminist rhetorical science studies shows that he shares this perspective's commitment to building a world that is accountable, ethical, and just. As such, his goal "align[s] with the social justice spirit of feminism" (Novotny, 2015, p. 62).

Bodies in the World

The majority of community rhetors embraced embodied vernacularity as an inventional resource. In addition to McConnell's hybrid strategy of drawing on institutional and vernacular discourses to craft authority, community members also established ethos through embodied vernacularity by focusing on bodies as they function in the world and acknowledging that bodies are sexual subjects. First, community rhetors concentrated on their bodies (or, in the case of RNs, clinicians, or nonprofit workers, their clients' bodies) as living, functioning bodies moving in the world. Community members used their everyday experiences of embodiment to provide context for the scientific data—context that often countered the institutional speakers' conclusions. For example, speakers provided graphic details of what exactly Truvada does to one's body—severe digestive issues, renal (kidney) issues, depression—and the probable effects of those "adverse events" on PrEP adherence. Nurses, physicians, nonprofit workers, and HIV-positive people themselves argued that people are not likely to take a drug that gives them daily diarrhea or risks severe kidney issues.

The data presented at the hearing indicated that in order for Truvada to be effective, it must be taken every day. Scientists' testimony tended to focus on *perfect* adherence, rather than *actual* adherence, arguing that the potential benefit of Truvada as PrEP—when taken perfectly—legitimized its approval. When they addressed nonadherence, it was often couched in statistical projections about Truvada's effectiveness *if* people were adherent. However, numerous community rhetors testified that daily adherence was unlikely if not impossible. Even without Truvada's side effects, community rhetors argued, perfect adherence to a daily Truvada regimen would be an impossibility for a variety of factors. According to community testimony, the most common reason for missing pills was a failure to remember to take them (U.S. FDA, 2012, pp. 318, 338). Additionally, "the high cost of the drug may make adherence even more difficult" (p. 327). Finally, community rhetors expressed concern that people with Truvada PrEP prescriptions may share their pills:

> I had a patient confide to me that her husband wouldn't come for treatment himself because of stigma, so she came and shared her medication with him. Knowing this happens with diagnosed patients, we logistically expect it even more in negative population. (pp. 338–339)

The content of community rhetors' testimonies often included vivid imagery of Truvada's effects on the body. For example, as RN "Maria Black" told the committee:

> One patient of mine. . . . took Truvada, which not only exposed him to problems with renal issues, but it pretty much guaranteed him daily diarrhea. And this is the kind where you need to know where the bathroom is all the time. And there were days when he simply didn't know where the bathroom was all the time and needed to be out, so he missed his doses. . . .
>
> I ask you, how do you justify that when you don't have HIV, when you aren't sick, and what you are taking will damage your kidneys and give you diarrhea every day? (pp. 265–266)

Black's testimony reveals a different way of thinking about bodies and evidence. Rather than seeing bodies—and particularly "at-risk" bodies—in the vacuum of clinical trial data, Black acknowledged the physicality of our bodies and the complexities of our lives. The focus of her testimony was not what happens on a plot or in a P-value but what happens to fleshy bodies in their actual lives. By positioning this bodily evidence as a counter to disembodied clinical trial data, Black enacted a vernacular of the body: expressive discourse asserting its alterity from the institutional through a rhetoric of embodiment. She used bodily specificities—the particulars of this patient's life—to make her case that Truvada's approval is not merited. Thinking about bodies functioning in the world in this way helps us remember that biomedical data exist beyond academic papers and posters, that data are gathered in the midst of the complexities of people's lives, schedules, and proximity to a toilet. These complexities matter when both considering ways to prevent HIV transmission and thinking about the relationship between science and society.

Thinking about bodies moving through the world and adopting a framework of embodied vernacularity can offer a different rhetorical perspective on the Truvada data. Embodied vernacularity—discourse that positions itself against an institution through a rhetoric of the body—shifts our rhetorical vision. Debra Hawhee (2011) notes that rhetorical vision "considers the visual work of rhetoric and language, the complex ways that words—oral or written—form perception" (p. 140). In other words, rhetorical vision means that what we say shapes what we see. Similarly, Haraway (1988) urges us to "plac[e] metaphorical reliance on a much maligned sensory system in feminist discourse: vision" (p. 581). Vision is inherently embodied. It relies on the fleshy body—eyes, brain—as well as metaphorical considerations of worldview and ontology.

Community rhetors used vision as a material and semiotic rhetorical resource when they spoke about PrEP's implementation. For example, when community members shared detailed concerns about side effects (kidney damage, diarrhea) and adherence (taking medication every day/as prescribed), they were employing the rhetorical strategy of *pro ommaton poiein*, or bringing-before-the-eyes. As Hawhee (2011) explains, "the net effect [of *pro ommaton poiein*] is an active, vivid engagement with the rhetor's words, such that hearers begin to engage in deliberative phantasia, leading to belief formation and decision making" (p. 159). Including vivid details in their testimonies allowed the audience (the advisory committee and the other people present at the hearing) to visualize what a daily Truvada regimen would be like. Community rhetors spoke about their experiences watching people take Truvada, witnessing the effects this drug has on people's bodies and psyches. In short, they presented an alternate vision of bodies that countered the institutional vision, which presumed that real-world adherence would match or exceed adherence during clinical trials. The institutional vision thus imagined PrEP rolling out in a social vacuum that failed to thoughtfully consider the social impacts on adherence.

At the Truvada hearing, then, two different kinds of bodies were evoked: an informatic, fungible body constructed by data on the one hand, and the individualized fleshy body on the other. A partial perspective, like that advocated by Haraway and deployed by the vernacular rhetors at the Truvada hearing, represents an alternative route to objectivity. The vernacular testimony, clearly and firmly situated in lived experience and fleshy bodies, had the potential to gain authority precisely by virtue of its limited location—that is, through specific and individualized embodiment. In the community testimony, partial perspective was able to gain authority through a theory of the flesh as articulated by Moraga and Anzaldúa.

The second way community members enacted an embodied vernacularity was by focusing on bodies as a sexual objects and subjects. Sex is often a complex negotiation of pleasure and risk, and condom use tends to be hailed as the ultimate HIV prevention strategy without acknowledgment that it can affect sexual pleasure, spontaneity, and trust (Higgins, Hoffman, & Dworkin, 2010). Many PrEP supporters acknowledged that HIV is often transmitted sexually, and they saw great promise in a prevention strategy that could allow people to move away from the incitement to use condoms "every time, all the time."[13] As "Jacob Potter" told the committee,

> what I'm asking for you is to allow people to be met *in the real locations of their lives*, to give them options and choices, to figure out prevention strategies that work for them, because condoms . . . aren't doing it. (U.S. FDA, 2012, p. 298, emphasis added)

In his testimony, Potter cites data showing a 48 percent increase in rates of new HIV infection among young black men who have sex with men. For Potter, condoms and existing prevention strategies are insufficient to meet people "in the real locations of their lives"—locations that are shaped by race, age, and sexual practices. For Potter, part of PrEP's great promise is that it enables us to acknowledge that condomless sex happens for complicated and varied reasons. Rather than vilify HIV-positive people for engaging in condomless sex, many supporters of PrEP argued that PrEP can better account for the embodied, situated sexual choices people make. Gayle Rubin (2011) articulates a sexual ethics that is divorced from conservative morality, and many PrEP supporters make their arguments from within such a feminist sexual ethic. Viewed in this way, visceral experiences of sexual pleasure become valid sites of authority. Many community rhetors challenged the committee to approve Truvada as PrEP precisely because it privileges sexual pleasure as a legitimate factor in sexual decision-making.

While the testimony offered by the community rhetors may not have enabled the committee to broaden their perception (a point I address further in the conclusion), perhaps it can still provide us access to epistemic paradigms rooted in bodies' knowledge. Even though the committee ultimately sided with the scientists and those community rhetors who supported PrEP, this should not be interpreted as a failure of embodied vernacularity, nor should it cause us to set aside embodied vernacular strategies. As feminist rhetoricians, we can heed J. Blake Scott's (2003) call that "the ultimate aim of a rhetorical-cultural study is ethical intervention, however tentative and local" (p. 229). Embodied vernacularity may have proven less persuasive than institutionalized scientific discourse, but it was an ethical act, one that sought to privilege the lived experiences of people with HIV. "Failed" rhetoric can still be valuable. We ought to remember that most activist movements "fail" many times before achieving victories. Perhaps the more community rhetors enact embodied vernacularity, the more we and they can open up space for that expertise to be valued.

Conclusion

In this chapter, I have shown how community activists embodied vernacularity during the FDA advisory committee meeting and demonstrated how community testimonies can reveal a great deal about the kinds of rhetorical strategies available to noninstitutional actors. The institutional legitimacy afforded to the invited scientific speakers—by virtue of their institutional affiliations and the power of scientific authority more broadly—enabled the scientists' particular and specific embodiments to fade into the background; their discourse, not their bodies or lived experiences, took center stage. Perhaps because this "God trick" (Haraway, 1988) was unavailable to the non-institutional rhetors, the community speakers sought creative ways to build ethos, including the strategy of embodied vernacularity. Importantly, this case study identifies not only the strategies employed but also these strategies' implications. It is these implications to which I turn now.

Unfortunately, vernacular voices were all but ignored by the committee. As I mentioned earlier, the committee approved Truvada for PrEP, even though the majority of community speakers opposed it. However, the real indication that the committee ignored the community comes from the voting portion of the meeting. Each committee member was asked to vote on several questions and explain their answer if they wanted to. Not one of the speakers—regardless of her or his vote—brought up community testimony as having influenced their decision. Scientific data and clinical experience were cited repeatedly, but the only mention of community came from "health disparities expert" Lauren Wood, who voted no, and explained:

> I think it's very important to let individuals know who came before during the public hearing to ... make sure that options, new options and new options in the toolbox, were available for black MSM [men who have sex with men], was one of the major considerations. (U.S. FDA, 2012, p. 514–515)

Wood notes the community's desire to have more prevention options, but she does not use that as justification for her no vote. Instead, she turns to the biomedical data:

> I have significant safety concerns because it's well-known that African Americans have an extreme disproportionate risk for end-stage renal disease, chronic kidney disease, and dialysis.
>
> This approval is based on data that was conducted in a total of 140 African American men for the MSM indication, 117 males from the iPrEx study and 23 males from the CDC study. (p. 515)

None of the voters mentioned the community members' objections to PrEP. None of the members seemed to view the community as having any sort of authority or legitimacy.

Although embodied vernacularity was not an effective rhetorical strategy in this particular rhetorical situation, that doesn't mean it lacks value or has nothing to teach us. Because embodied rhetoric enacts an alternate ethos, it can be an effective rhetorical strategy for vernacular rhetors seeking to negotiate an asymmetrical power relationship and work toward a place of equal authority. Perhaps the testimony of the community rhetors engendered an affective response for the audience. Hearing vivid details of side effects, or hearing about the importance of sex and sexuality, may have caused the audience of other community members and scientists to be emotionally moved. As Jenell Johnson (2014) notes, "to acknowledge one's emotion in scientific or medical discourse is thus a risky endeavor, for calling attention to embodiment, partiality, and attachment challenges medicine's identity as an applied science, which draws its authority from norms of universalism and disinterestedness" (p. 21). However, although this move is risky, it is also valuable and can serve to equalize the playing field, as it were, between institutional and vernacular actors. Affect can help resituate knowledge in the body.

Further, and more importantly, what the community testimony shows is that bodies move in the world, they have schedules and commitments that might affect pill adherence, struggle with side effects of pharmaceutical regimens, and seek pleasure and have sex. The scientific presentations and discussions at the Truvada hearing minimized these embodied experiences in favor of aggregated statistical data, resulting in a scientific discourse that is, in many ways, incomplete. If institutional actors were to pay attention to the embodied vernacularity put forth by community members, we may end up with science that is better able to account for and respond to complex human phenomena like adherence.

Reading vernacular voices in the deep context from which they emerge also allows us to do rhetoric differently, more ethically. Jessica Enoch's (2005) historiographic approach to studying the once-ignored testimonies of marginalized rhetors demonstrates that recovering these testimonies is an important scholarly *and* political project. Similarly, in this project, recovering and devoting attention to the "failed" rhetorics of community members who opposed Truvada as PrEP can show us how we can recuperate moments of "failed" rhetoric as valuable objects of study, not for their immediate impact but for their intent and role in generating future effects.

Community rhetors can (and should) continue to draw on their fleshy bodies as they make their arguments, pushing to expand the rhetorical terrain available to interlocutors. AIDS activism has a long history of using the body as argument in activist settings (DeLuca, 1999), and similar strategies may come to be effective in institutional settings as well, should they be used often enough. Rhetoric scholars can notice these moments of embodied vernacularity and rethink what "failing" means—were the community rhetors at the Truvada hearing successful at expanding the argumentative sphere, even if they failed to convince their audience? The concept of embodied vernacularity developed in this chapter reveals the ways that power asymmetries manifest in deliberative settings, allowing us to understand not only the importance of a rhetor's words but also her or his particular speaking, fleshy body. Perhaps too—and this may be the greatest hope of embodied vernacularity—listening to bodies can result in better scientific practices.

Notes

The author wishes to thank Jenell Johnson, Sara McKinnon, Kadin Henningsen, and Whitney Gent for their careful readings of this chapter, as well as the editors and reviewers for their valuable feedback.

1. The clinical research data from several PrEP trials show two important findings (Grant et al., 2010). First, when taken daily as PrEP, Truvada (the drug being studied) was over 90 percent effective at preventing HIV transmission between male sex partners when taken as prescribed. Second, the same data also showed extremely low adherence, meaning that the trial participants were often not taking the drug as prescribed. Overall, then, Truvada was only about 44 percent effective as prevention, since participants were not actually taking the drug daily.

2. Activists and advocates have long challenged the conflation of "healthy" and "HIV-negative," since people with HIV are often quite healthy, and the absence of HIV does not necessarily make someone healthy. This persistent and troubling slippage between health and seronegativity has real-world implications for participation in clinical research (and also for federal policies derived from clinical research data). Although there is variability across both individual research studies and areas of research, the designation "healthy volunteers" in clinical research and care still generally excludes people with HIV.

3. Gilead is one of the largest pharmaceutical companies in the United States, with product sales of $10.8 billion in 2013 (Gilead Sciences, 2014). Several commonly prescribed HIV drugs are Gilead products, and more HIV drugs are in the company's development pipeline.

4. For lengthier discussions of cultural trust in science, see, for example, Bourdieu, 2004; Hess, 1997; Shapin and Schaffer, 1989.

5. *Serodiscordant* refers to people who have different HIV statuses.

6. See, for example, Antelius, 2009; Dolmage, 2009; Hawhee, 2011; Jordan, 2004; MacCannell and Zakarin, 1994; McKerrow, 1998; Patterson and Corning, 1997.

7. Although many of the people who testified during the community portion of the meeting were doctors, nurses, and heads of nonprofit organizations, and

could therefore be described as institutional rhetors, they spoke in response (and in opposition) to the FDA. If we apply Howard's conceptual framework, we can describe them instead as vernacular rhetors.

8. When a community member's testimony reached the three-minute mark, the chair of the committee had their microphone turned off, even if the speaker was midsentence. This did not happen to any of the scientific speakers.

9. Since the 2012 decision to approve PrEP, the discursive tide has shifted dramatically. Outspoken critics of PrEP, such as the AIDS Healthcare Foundation, have been met with backlash from members of the gay community, AIDS activists, and AIDS health-care workers. Further, more data have become available (AIDSMeds, 2014), suggesting that intermittent Truvada dosing works as well as daily dosing (for example, taking Truvada for just a few days before and after unprotected sex with an HIV-positive partner). Despite the broader vernacular shift toward pro-PrEP discourse since Truvada's approval, the 2012 community testimony against PrEP remains significant, if not for the specific arguments made, then for the argumentative strategies employed.

10. I have chosen to use pseudonyms for the community speakers throughout this chapter. While community members' real names are included in the official public transcript, I wish to protect the privacy of the community members here, particularly in cases where individuals disclose their HIV status. There are different stakes for those speaking on behalf of institutions and those speaking from personal experiences as advocates and HIV-positive individuals.

11. McConnell was number eighteen on the list of community speakers, but several speakers were not present at the hearing, so he actually spoke fourteenth.

12. Since Truvada's approval, numerous community-based activist and advocacy groups have echoed McConnell's concerns about ethics and taken stances on gender, racial, and class-based inequalities in access to Truvada as PrEP (for example, Black AIDS Institute, 2016; U.S. Women and PrEP Working Group, 2013).

13. This is a safe sex mantra that has gained popularity among public health workers and some members of LGBT communities over the last thirty years.

References

Ahmed, S. (2006). *Queer phenomenology: Orientations, objects, others*. Durham, NC: Duke University Press.

AIDSMeds. (2014, October 29). High efficacy halts placebo phase of intermittent PrEP study. *POZ*. Accessed 27 December 2014. Retrieved from https://www.poz.com/article/PrEP-IPERGAY-26344-9591

Antelius, E. (2009). Whose body is it anyway? Verbalization, embodiment, and the creation of narratives. *Health, 13*(3), 361–379.

Black AIDS Institute. (2016). Black lives matter: What's PrEP got to do with it? *Black AIDS*. Accessed 10 August 2016. Retrieved from https://www.blackaids.org/images/reports/16%20prep%20report.pdf

Bourdieu, P. (2004). *Science of science and reflexivity*. Chicago, IL: University of Chicago Press.

Brouwer, D. C. (2001). ACT-ing UP in congressional hearings. In R. Asen & D. C. Brouwer (Eds.), *Counterpublics and the state* (pp. 87–110). Albany: State University of New York Press.

Butler, J. (1993). *Bodies that matter: On the discursive limits of "sex."* New York, NY: Routledge.

Centers for Disease Control and Prevention. (2017, June). HIV in the United States: At a glance. *CDC*. Accessed 6 July 2017. Retrieved from https://www.cdc.gov/hiv/statistics/overview/ataglance.html

Chávez, K. R. (2009). Embodied translation: Dominant discourse and communication with migrant bodies-as-text. *Howard Journal of Communications, 20*(1), 18–36.

Clare, E. (2003). Gawking, gaping, staring. *GLQ, 9*(1–2), 257–261.

Clough, S. (2013). Pragmatism and embodiment as resources for feminist interventions in science. *Contemporary Pragmatism, 10*(2), 121–134.

Clughen, L. (2014). "Embodied writing support": The importance of the body in engaging students with writing. *Journal of Writing in Creative Practice, 7*(2), 283–300.

Condit, C. (1996). How bad science stays that way: Brain sex, demarcation, and the status of truth in the rhetoric of science. *Rhetoric Society Quarterly, 26*(4), 83–109.

Crimp, D. (1987). How to have promiscuity in an epidemic. *October, 43*, 237–271.

DeLuca, K. M. (1999). Unruly arguments: The body rhetoric of Earth First!, ACT UP, and Queer Nation. *Argumentation and Advocacy, 36*(1), 9–21.

Dolmage, J. (2009). Metis, mêtis, mestiza, medusa: Rhetorical bodies across rhetorical traditions. *Rhetoric Review, 28*(1), 1–28.

Enoch, J. (2005). Survival stories: Feminist historiographic approaches to Chicana rhetorics of sterilization abuse. *Rhetoric Society Quarterly, 35*(3), 5–30.

Epstein, S. (1995). The construction of lay expertise: AIDS activism and the forging of credibility in the reform of clinical trials. *Science, Technology, and Human Values, 20*(4), 408–437.

Gilead Sciences. (2014). Gilead Sciences announces fourth quarter and full year 2013 financial results. *Gilead*. Accessed 27 December 2014. Retrieved from http://www.gilead.com/news/press-releases/2014/2/gilead-sciences-announces-fourth-quarter-and-full-year-2013-financial-results

Grant, R. M., et al. (2010). Preexposure chemoprophylaxis for HIV prevention in men who have sex with men. *New England Journal of Medicine, 363*(27), 2587–2599.

Haraway, D. (1988). Situated knowledges: The science question in feminism and the privilege of partial perspective. *Feminist Studies, 14*(3), 575–599.

Hauser, G. A. (1999). *Vernacular voices: The rhetoric of publics and public spheres.* Columbia: University of South Carolina Press.

Hawhee, D. (2011). Looking into Aristotle's eyes: Toward a theory of rhetorical vision. *Advances in the History of Rhetoric, 14*(2), 139–165.

Hess, D. J. (1997). *Science studies: An advanced introduction.* New York: New York University Press.

Higgins, J. A., Hoffman, S., & Dworkin, S. L. (2010). Rethinking gender, heterosexual men, and women's vulnerability to HIV/AIDS. *American Journal of Public Health, 100*(3), 435–445.

hooks, b. (2000). *Feminism is for everybody: Passionate politics.* Cambridge, MA: South End Press.

Howard, R. G. (2010). The vernacular mode: Locating the non-institutional in the practice of citizenship. In D. C. Brouwer & R. Asen (Eds.), *Public modalities: Rhetoric, culture, media, and the shape of public life* (pp. 240–262). Tuscaloosa: University of Alabama Press.

Johnson, J. (2014). *American lobotomy: A rhetorical history.* Ann Arbor: University of Michigan Press.

Jordan, J. W. (2004). The rhetorical limits of the "plastic body." *Quarterly Journal of Speech, 90*(3), 327–358.

Kagan, J. M., et al. (2012). Community-researcher partnerships at NIAID HIV/AIDS clinical trials sites: Insights for evaluation and enhancement. *Progress in Community Health Partnerships, 6*(3), 311–320.

MacCannell, J. F., & Zakarin, L. (Eds.). (1994). *Thinking bodies.* Stanford, CA: Stanford University Press.

McKerrow, R. E. (1998). Corporeality and cultural rhetoric: A site for rhetoric's future. *Southern Communication Journal, 63*(4), 315–328.

Moraga, C., & Anzaldúa, G. (Eds.). (1983). *This bridge called my back: Writings by radical women of color* (2nd ed.). New York, NY: Kitchen Table—Women of Color Press.

Novotny, M. (2015). ReVITALize gynecology: Reimagining apparent feminism's methodology in participatory health intervention projects. *Communication Design Quarterly, 3*(4), 61–74.

Ono, K. A., & Sloop, J. M. (1995). The critique of vernacular discourse. *Communication Monographs, 62*(1), 19–46.

Patterson, R., & Corning, G. (1997). Researching the body: An annotated bibliography for rhetoric. *Rhetoric Society Quarterly, 27*(3), 5–29.

Price, M. (2015). The bodymind problem and the possibilities of pain. *Hypatia, 30*(1), 268–284.

Reeves, C. (1992). Owning a virus: The rhetoric of scientific discovery accounts. *Rhetoric Review, 10*(2), 321–336.

Reeves, C. (1998). Rhetoric and the AIDS virus hunt. *Quarterly Journal of Speech, 84*(1), 1–22.

Rubin, G. S. (2011). Thinking sex: Notes for a radical theory of the politics of sexuality. In *Deviations: A Gayle Rubin reader* (pp. 137–181). Durham, NC: Duke University Press.

Scott, J. B. (2003). *Risky rhetoric: AIDS and the cultural practices of HIV testing.* Carbondale: Southern Illinois University Press.

Shapin, S., and Schaffer, S. (1989). *Leviathan and the air-pump: Hobbes, Boyle, and the experimental life.* Princeton, NJ: Princeton University Press.

Shome, R. (2003). Space matters: The power and practice of space. *Communication Theory, 13*(1), 39–56.

Treichler, P. A. (1999). *How to have theory in an epidemic: Cultural chronicles of AIDS.* Durham, NC: Duke University Press.

U.S. Food and Drug Administration, Center for Drug Evaluation and Research. (2012). Transcript of Antiviral Drugs Advisory Committee meeting, May 10, 2012. *FDA.* Accessed 19 October 2012. Retrieved from http://www.fda.gov/downloads/AdvisoryCommittees/CommitteesMeetingMaterials/Drugs/AntiviralDrugsAdvisoryCommittee/UCM311519.pdf

U.S. Women and PrEP Working Group. (2013, March 4). Working group on U.S. women and PrEP statement. *AVAC.* Accessed 2 August 2016. Retrieved from http://www.avac.org/sites/default/files/u44/Women_and_PrEP_Statement_March2013.pdf

Willey, A. (2016). Biopossibility: A queer feminist materialist science studies manifesto, with special reference to the question of monogamous behavior. *Signs, 41*(3), 553–577.

8.
Becoming-Thinking Otherwise, Rhetorically

Amanda K. Booher and Julie Jung

Striving to develop and model ways of doing feminist posthumanist research in rhetorics of science, many of our contributors draw on scholarship in feminist new materialism. In this final chapter, we aim to reverse that trajectory by demonstrating how research in feminist rhetorics can extend work in feminist new materialism. In so doing, we hope to model the kind of cross-disciplinary reciprocity for which we advocate.

We begin by reviewing a key critique of feminist new materialism that has important implications for feminist rhetorics: its inability to guide concrete political action in the work of social justice. We then respond to this critique by discussing how a theory of feminist new materialist rhetorical practice, which effects changes in habits of domination, can productively link the political need to challenge sustained systemic oppressions with new materialism's future-oriented focus on ontological becomings. We conclude with analyses of research in activist rhetorics that demonstrate how scholarship in rhetoric studies can contribute to feminist science studies (FSS) in general and feminist new materialism specifically.

Feminist New Materialism: A Critique

In "Ontologized Agency and Political Critique," Bonnie Washick and Elizabeth Wingrove (2015) review scholarship in feminist new materialism, focusing primarily on Karen Barad's theory of agential realism and Jane Bennett's conceptualization of assemblagic agency, to question the degree to which its posthumanist ontology can "illuminate any particular form of action—or any privileged mode of 'intra-action'—that corresponds to the cooperative, conflictual, agonistic, and/or deliberative relations variously associated with political engagement" (p. 66).[1] The authors are specifically concerned with how new materialist accounts of power shift focus "away from structural constraints and toward 'agentic possibilities that do not stand still'" (p. 70). In other words, by focusing on the flux and vitality of matter, which holds the potential that future worlds might materialize otherwise, feminist new materialism fails to contend with the complex ways in which *already* existing materializations—enacted in durable social structures that sustain injustice—have a hand in delimiting what the world might otherwise become. With reference to Barad specifically, the authors contend that her "emphasis on 'the mutual constitution of agencies' ... make[s] it difficult to name and so hold in view the continuities, durabilities and often monotonous predictabilities that characterize systems of power asymmetry (such as capitalism, patriarchy, racism)" (pp. 65–66). For feminist rhetoricians committed to taking political action in the service of social justice, Washick and Wingrove's critique should give us pause, for it asks us to take seriously how a focus on our shared potential for enacting different worlds downplays the ways in which bodies "are *differentially* constrained" by "the systematic reproduction of inequalities" (pp. 69, 78).

The authors clarify that they recognize feminist new materialism as politically motivated, as "deeply embedded in the imperatives of critique arising within the current historical context" (p. 77). At issue here is the question of how to convert the abstractions of its relational ontology into concrete activist practices designed to rectify social injustices. In what follows, we argue that a feminist rhetorical conceptualization of habit can help facilitate this conversion.

Habits Are Practices

In their study of feminist rhetorical practices, Jacqueline Jones Royster and Gesa E. Kirsch (2012) read across feminist rhetorical scholarship to observe what teacher-scholars are actually doing (their practices), identify patterns among these doings, and then frame these doings through methodological concepts (for example, critical imagination, strategic contemplation) that can guide future research practices. The authors implicitly render a distinction between practice and habit when they argue that

> practices in feminist rhetorical studies have broken through the habitual expectations for rhetorical studies to be overwhelmingly about men and male-dominated arenas, with the consequence of creating volatility in research and practice, tectonic shifts on the rhetorical landscape. We argue further that this shifting process has not been arbitrary or mystical

but the result of specific strategies and methodologies that are forming now-coherent patterns for well-grounded action. (p. 17)

Their use of "practice" contrasts with "habit" in the sense that the latter signifies a particular kind of practice: an expected way of doing things, where in that expectation resides a nonconscious agreement that these ways of doing are both natural and good. Such a conceptualization aligns with theories that posit habits as "automatic, preconscious, predictable, coordinated, consistent activities" (Hansen, 2013, p. 71). As Royster and Kirsch (2012) remind us, it had become a habit in rhetoric studies to see only White male public orators as rhetors who "count," but this habit was disrupted by the "rescue, recovery, [and] (re)inscription" (p. 14) projects of feminist rhetorical historiographers. In disrupting the habit, these scholars called attention to the need to do things otherwise. These "doings otherwise" Royster and Kirsch characterize as "feminist rhetorical practices": "deliberate and self-conscious" (p. 17) attempts to disrupt expectations and thus researchers' tacit agreement to continue doing things the same way. This specification of practice differs from its traditional definition as "a set of tasks one repeatedly undertakes to acquire or improve a skill" (Boyle, 2016, p. 544). Instead, we can understand feminist rhetorical practices as repeated actions that disagree with and thus challenge habits of domination and exclusion.

To be sure, there are other ways to understand habit. Indeed, in his theory of posthuman practice, Casey Boyle (2016) posits that we in the field would be well served "to focus not only on what habits to encourage but also how existing habits can be made differently productive" (p. 550). But the articulation we advocate for here is designed to address Washick and Wingrove's critique: we can understand normalizing habits as nonconscious human actions that enact enduring inequalities *and* possibilities for concrete political change. Such an articulation also presses on the posthumanist position that public rhetoric is less about solving problems and more about forging associations that become the conditions of possibility for "the *emergence* of solutions" (Hawk, 2011, p. 90). Specifically, it asks us to recognize that problems do indeed now exist, that these problems have complex histories that render them durable and systemic, and that assemblages capable of producing emergent solutions to existing problems must include these histories among their constitutive elements.

Catherine L. Langford and Monteré Speight's (2015) rhetorical analysis of #BlackLivesMatter serves as an effective site to think through the affordances of this conceptualization of normalizing habit. In their article, Langford and Speight discuss how the movement "draw[s] attention to the habitual violence against Blacks in America" (p. 78). In describing this violence as "habitual," the authors gesture toward its being both historically enduring and systemic: it is enabled by a dominant public pedagogy that, "through slavery, black codes, Jim Crow, lynchings, poverty, and shootings[,] coaches White society that Black lives do not matter" (p. 79).[2] Their project also serves as a potent reminder that "habit is always habit in a world" (Engman & Cranford, 2016, p. 40); thus, paying attention to habits means attending to the ways in which specific worlds and particular kinds of habits are co-constituted. A world habituated to a pedagogy of racism constitutes and is constituted by repeated acts of violence toward Blacks. This is so in part because, as Linda Martín Alcoff (2006), drawing on Maurice Merleau-Ponty, explains,

> our experience of habitual perceptions is so attenuated as to skip the stage of conscious interpretation and intent. Indeed, interpretation is the wrong word here: we are simply perceiving.... A fear of African Americans or a condescension toward Latinos is seen as simple perception of the real, justified by the nature of things in themselves.... [T]he [racial] profiler does not understand him or herself to be using judgment at all but simply perceiving danger. (pp. 188, 197)

In their theorizing of perception, new materialist theorists affirm Alcoff's phenomenological reading, which posits that existing structures of oppression and relations of power are manifested in embodied habits that sustain the status quo. As Diana Coole and Samantha Frost (2010) explain, histories of disciplining inscribed in daily, repetitive actions "become sedimented at a corporeal level, where they are repeated as habits or taken for granted know-how: lodged in the bodily memory that Bourdieu calls *habitus*" (34; see also Connolly, 2010). Indeed, it is precisely this idea of sedimentation that is crucial to Barad's (2003) theory of intra-action, which she describes as "causally constraining nondeterministic enactments through which matter-in-the process-of-becoming is sedimented out and enfolded in further materializations" (p. 823). Talking specifically about Baradian intra-action, gender theorist Gill Jagger (2015) explains:

> The concept of sedimentation is significant..., as reality in agential realist terms consists of the sedimentation of particular intra-actions and boundary-making practices that have produced intelligible configurations (or materializations). Sedimentation thus indicates an ongoing process of configuration and reconfiguration, involving both human and nonhuman agencies, a process that constitutes reality and yet is open to change. (p. 330)

When normalizing habits are posited as enactments of sedimented power relations, we have the means to coherently merge Washick and Wingrove's call for a new materialist theory that accounts for historically sustained structures of oppression—which *have already* materialized through certain kinds of intra-actions—with its ontology of becoming, which recognizes those structures as constraining possibilities for future intra-actions but not determinative of them.

Habits Are Persuadable

From a posthumanist perspective, the repeated nonconscious actions associated with embodied habits cannot be attributed solely to the individual human body that acts. Rather, they are ways of doing things that emerge from systems of relation that include the nonhuman. As Boyle (2016) explains, all practices are "ecological" and "emerge from particular situations, distributed across a variety of material relationships, and are temporally contingent" (p. 541). On this view, our practices are never our "own" per se, yet we are part of the system from which practices emerge. It follows, then, that effecting a change in habituated practices necessarily involves changing the terms of relation through which those habits emerge. Habits, Boyle explains, are "firm but flexible, positioned but persuadable" (p. 550). Possibilities for action open up when we recognize that habits are persuadable. From a posthumanist rhetorical perspective, then, we might focus on trying to persuade habits instead of people. When we help to enact changes in ways of relating among elements in a system, other ways of doing things become possible.

Of course, this is precisely what existing feminist rhetorical practices, such as those advocated by Royster and Kirsch, aim to do: disrupt disciplinary habits by changing the ways teacher-scholars relate to their subjects of inquiry. What a posthumanist orientation adds to the conversation is a willingness to pay attention to how a diverse array of elements, including the nonhuman, might participate in such refigurings. Here we draw on and extend Boyle's posthumanist understanding of rhetorical practice, which is "less concerned about conscious awareness of being embedded and more concerned with inventing techniques, many of which operate on nonconscious levels with which we exercise that embeddedness" (p. 538). Retaining Boyle's emphasis on inventing techniques that enact relations, we also ask: for those of us who seek not only to understand how change happens but also to influence which changes come into being, what might we gain if we regard nonhuman objects as rhetorical actants capable of disrupting habits that normalize injustice?

We see evidence of this gain throughout this collection. Jen Talbot's study of the ultrasound and the Display Requirement in "Flat Ontologies and Everyday Feminisms: Revisiting Personhood and Fetal Ultrasound Imaging" inverts our question slightly, considering how nonhuman objects reinforce injustice in laws limiting abortion rights. For instance, she notes that ultrasounds function as rhetorical actants "to allow the mother to experience the fetus visually and aurally, to allow people other than the mother to experience the ontological reality of the fetus without the mediation of the mother, and to imbue the fetus with characteristics." These allowances then work to help establish legal personhood, and thus legal rights, for the fetus over those of the mother.

In "'The Inconvenience of Meeting You': Rereading Non/Compliance, Enabling Care," Catherine Gouge argues for a "kairology of care" to disrupt reliance on the notion of sole human agency in questions of medical compliance. Her kairology recognizes that those who might be considered willfully noncompliant are instead "opportunistically working with the constantly changing affordances and constraints of their situated, embodied experience." Thus, she suggests that a more ethical approach to understanding (non)compliance demands consideration of not only the patients but also the nonhuman "affordances and constraints" that are actants in these patients' behaviors. In "Embodied Vernacularity at the FDA: Feminism, Epistemic Authority, and Biomedical Activism," Liz Barr similarly argues for the need to consider people's behaviors as complexly enmeshed with the nonhuman, a complexity that traditional means for assessing the efficacy and safety of drugs like Truvada fail to capture. As Barr argues, we must remember that

> biomedical data exist beyond academic papers and posters, that data are gathered in the midst of the complexities of people's lives, schedules, and proximity to a toilet. These complexities matter when both considering ways to prevent HIV transmission and thinking about the relationship between science and society.

In this way, Barr's project calls on us to consider embodied phenomena like diarrhea not simply as side effects associated with drug use but rather as rhetorical actants capable of disrupting, even momentarily, the habit of privileging the authority of disembodied scientific data.[3]

Jennifer Bay, in "Mattering Gender: Technical Communication and Human Materiality," attempts to influence change by exploring what new materialist theory can bring to our students, particularly women in technical communication who are entering "stereotypically masculinist" industries. She asks us to consider what and how we teach about/with technology:

> What would it mean to have students see technological artifacts as agents rather than as things to be conquered or mastered? Would that give them a different sense of the world? Rather than working against things, could they work with them? And how would such knowledge and experience play out in coworking scenarios or any other kind of distributed work, rife as they may be with gendered power differentials?

These questions themselves work to disrupt our habits, as do the considerations she advocates. Among these habits Bay includes unquestioned research practices that can render invisible the very phenomena we seek to understand, such as human users' engagements with technologies. Given that this is a primary concern in Technical Communication, Bay advocates that scholars in the field develop methodologies informed by feminist new materialism, which "takes both human beings and objects as equally productive agents" and thus more effectively "allow[s] for our relationships with material artifacts" and "certain complex materialities of gendered existence to show up in our research."

In "How Good Brain Science Gets That Way: Reclaiming the Scientific Study of Sexed and Gendered Brains," Jordynn Jack similarly calls on scholars in feminist rhetorical science studies (FRSS) to rethink scholarly habits that restrict our capacity as rhetoricians to effect change. Specifically, Jack asks us to see ourselves as having a key role in producing science, rather just critiquing it. By focusing on "the experimental situation itself"—interactions among humans, concepts, measuring apparatuses, and design protocols, where "even small items like a form" or "rewards given for participation" shape outcomes—we can better understand how gender stereotypes and gender itself *emerge* in and through scientific research and thus intervene more effectively in their problematic re/production. Such intervention, however, requires that we move beyond our disciplinary comfort zones and recognize that our "work begins long before an experimental result is published."

Differently considering the ethics of medically related policies, Daniel J. Card, Molly M. Kessler, and S. Scott Graham, in "Representing without Representation: A Feminist New Materialist Exploration of Federal Pharmaceutical Policy," critique the U.S. Food and Drug Administration's effort to "ensure that patients and consumers are adequately represented in policy forums traditionally dominated by biomedical researchers and health-care practitioners." Noting that this approach of representational justice is not borne out in the transcripts of such meetings, they argue that "an ethical health policy forum must be engineered so as to include . . . (perhaps more importantly [than the testimony of people]) the testimony of spaces, bodies, and practices." They advocate for a "politics of what" approach, determining that considering the emergence of "primary ontologies"—in their study, lab, home, clinic, and market—can more effectively disrupt the potential injustices of representation alone.

While the earlier discussion of contributor chapters considers how nonhuman elements participate in disrupting habituated practices, we also believe there is more feminist posthumanist scholars can learn from research in activist rhetorics that are not explicitly posthumanist. This research, we argue, suggests ways of inventing specific practices that in changing how elements relate within a system disrupt the smooth emergence of habits of domination. In support of this argument, we return to Langford and Speight's (2015) rhetorical analysis of the #BlackLivesMatter movement, where the authors explain how the movement has helped to dislodge sedimented perceptual habits that authorize racial violence by "re-script[ing] the Black body as valuable" (p. 86). In keeping the violence "perpetuated against Black bodies within the public consciousness" and rendering that violence "a newsworthy event," the movement challenges its status as both "normal and unremarkable" (p. 87). By disrupting the dominant pedagogy, the movement revises how humans relate to another, thereby helping to bring into White consciousness the habit of accepting police actions without question.

A posthumanist rhetorical perspective calls on us to extend this analysis by focusing on the hashtag itself as a rhetorical actant: in circulating through social media and "enter[ing] into relation with other things (human or nonhuman)," it became able "to induce change in thought, feeling, and action; organize and maintain collective formation; [and] exert power" (Gries, 2015, p. 11). These relations include not only the hashtag's interactions with human users (how it was taken up and circulated) but also those users' technologies of circulation—iPhones, Facebook, Tumblr, Twitter, and so on.

We can also regard the hashtag's rhetorical actancy as emerging in part from the "grief and sorrow" experienced by its cofounders, Alicia Garza, Patrisse Cullors, and Opal Tometi, after George Zimmerman's acquittal in the shooting death of Trayvon Martin (Langford & Speight, 2015, p. 82). These embodied experiences themselves emerged in part from dominant pedagogies that authorize racial violence despite a long history of African American activist efforts dedicated to ending that violence. In short, three women, motivated by a desire to affirm Black lives, made something. Not intended to change people's minds in a traditional rhetorical sense, their creation entered into and altered a system of relating such that the dominant habit of accepting police officers' use of deadly force in their encounters with unarmed Black persons could no longer emerge without question as being both natural and right.

Objects Are Phenomena

The earlier discussion of the hashtag's *rhetorical actancy* in the context of the #BlackLivesMatter movement—its potential to alter relations and thus disrupt habits of domination—is indebted to Laurie Gries's (2015) articulation of the concept, which "insist[s] that collective life is constituted by a multiplicity of circulating entities that are mutually influencing each other and bending space as a consequence of their divergent activities" (p. 75). When one of those entities is an object, we might be tempted to conclude that rhetorical actancy is something that object *has*, which would contradict new materialist assumptions about the vitality and ongoing materialization of matter. Describing rhetorical actancy using a new materialist framework, Gries explains that "an object *becomes* rhetorical," and it does so, "in part, as a consequence of [its] emergent ability to mobilize various entities into relation, help materialize change, and thus reassemble collective existence" (p. 13, emphasis added). To delve further into this process of becoming as it relates to our earlier discussion of habit, we return to Barad, whose theories, while enormously influential, can in their abstraction occlude their potential to serve feminist politics in concrete ways.

As both Jen Talbot and Jennifer Bay discuss in their respective chapters, Barad (2007) argues that reality is constituted not by fixed, inert objects but by phenomena, "ontologically primitive relations—relations without preexisting relata" (p. 139). As Barad (2003) herself explains,

> it is through specific agential intra-actions that the boundaries and properties of the "components" of phenomena become determinate and that particular embodied concepts become meaningful. A specific intra-action ... enacts an *agential cut* (in contrast to the Cartesian cut—an inherent distinction—between subject and object) effecting a separation between "subject" and "object." ... [R]elata do not preexist relations; rather, relata-within-phenomena emerge through specific intra-actions. (p. 815)

This shift from "object" to "phenomenon" emphasizes that what we recognize as a discrete and boundaried object emerges through specific intra-actions that come to matter—that is, they both materialize and are rendered intelligible—in a specific but nondeterminative way. The phenomenon of unarmed Black persons being shot by police materializes through co-constitutive intra-actions among elements such as histories of race relations, practices of institutionalized racism, guns, police uniforms, discourses of criminality, police training—and, most significantly given our purposes here, the White majority's nonconscious agreement to accept as reasonable police officers' violent treatment of Black bodies. Through Barad's theory, what were previously understood to be independent and separate *parts* of a phenomenon—*its cause* (the threatening Black body, especially the Black male body) and *effect* (the police officer acting reasonably); *the objects* (criminals) and *the subjects* (police officers acting in our best interests)—are now understood as emergent distinctions rendered intelligible by the very intra-actions through which the body left bleeding on the road came into being. Thus, while only some Black individuals are shot dead by police, we *all* intra-act with the phenomenon and thus help sustain the distinctions that make it possible for the White majority to interpret such acts of violence as reasonable, even good. Yet significantly these intra-actions produce a contingent rather than an essential or determined difference; thus, we are all capable of intervening: we can remake boundaries and meanings of difference by enacting and helping others to enact alternative practices of material-discursive entanglement.

The hashtag #BlackLivesMatter enacted one such alternative by altering habits of intra-action. Not an inert object that was simply inserted into a preexisting system of relating, the hashtag emerged as a rhetorical actant through and participated agentially in intra-actions that created different distinctions. In making new cuts, these intra-actions refigured the intelligibility of cause/effect: the dead body in the road is caused by the actions of police officers whose perception of threat is inextricable from racism; the effect is a Black man murdered by police. From a Baradian perspective, then, the hashtag doesn't simply disrupt the normalizing habit because it changes the system of relating through which the habit emerges; rather, the hashtag is intra-acting in ways such that the habit itself is in the process of *becoming something else*. Simply put: new materialist rhetorical practices don't disrupt habits; they enact possibilities for creating new ones.

Feminist New Materialist Rhetorical Practices Are Concrete Political Actions

What emerges from the previous discussion—which blends posthumanist and activist rhetorical theories of habit and object with Royster and Kirsch's articulation of feminist rhetorical practices—is a theory of *feminist new materialist rhetorical practice* that offers specific tactics for engaging in concrete political action. These tactics are as follows: identify habits of exclusion and domination; make a material thing that renounces those habits; share that thing with others; and then pay attention to how the thing *as a phenomenon* becomes rhetorical: What new collectives does it help call into being? What changes does it participate in materializing? What new things does it ask us to make?

To demonstrate how these tactics have effected change in the scene of FRSS, we turn now to Susan Wells's *"Our Bodies, Ourselves" and the Work of Writing* (2010), a complex study of the book *Our Bodies, Ourselves* (*OBOS*) as it was produced by the Boston Women's Health Collective from 1973–1984.[4] Given the important place of *OBOS* in FSS origin stories, Wells's book makes a vital contribution to scholarly histories of FSS (though the publisher markets it as belonging in "Literary Studies/Women's Studies"). It is also a *historical* study, and feminist histories of science constitute a major area of inquiry in FSS to which feminist rhetoricians have much to contribute.[5] Most important, given our purposes, is how Wells's book demonstrates the affordances of scholarly projects that take as their objects of inquiry feminist activist rhetorics. Although this book was published in 2010, before our work here, and a comprehensive discussion of it is beyond the scope of our chapter, Wells's project clearly demonstrates the benefits of thinking about feminist texts as phenomena capable of intervening in hegemonic scientific practices.

According to Wells, members of Boston Women's Health Collective were concerned with, among other things, habitual practices of medical mystification, which alienated women from their own bodies and constructed them as passive recipients of physician expertise. As such, the collective sought to produce a text that provided readers with "an accessible critical analysis of scientific medicine and a new way of thinking about [their] own bod[ies]" (p. 61). In a conventional rhetorical analysis, a rhetorician would proceed by analyzing specific strategies the collective used to appeal to its target audience in order to accomplish its purpose. Such an approach would necessarily focus on the collective's intentional language choices. Wells, however, extends her analysis to consider the collective's language use in relation to the book's physical design and readers' physical use of it as a tool for bodily self-investigation. Specifically, Wells explains how the collective juxtaposed personal narratives that mimicked intimate conversation alongside complex medical research such that readers were "encouraged . . . to move into the book" and become participants in the construction of shared knowledge (p. 100). Such discursive strategies were coupled with suggestions for organizing and using the book in its material form. As Wells explains in regards to the first 1971 edition,

> they printed the book cheaply on newsprint. . . . Readers who wanted to put the pages in a notebook were instructed to punch holes in the margin, to slit the binding thread and the back of each booklet with a razor blade, and to put the book into a binder. Health classes could take it apart and distribute each of the four booklets separately. The book looked handmade, and in that first edition, readers were encouraged to participate in making it over themselves[,] . . . to shape [their] cop[ies] to [their] own needs. (p. 101)

One such need imagined by the collective was that the reader would use the book as a prop to overcome her own bodily alienation: she, "book in hand, would look into a mirror at her genitals" (p. 100). With the book, the reader became a bookmaker and an embodied knower; with the reader, the book materialized as an object motivating and enabling embodied knowing. New ways of relating disrupted habits of medical mystification and thus emerged as agencies of change.

When *OBOS* is conceptualized not just as an object but as a phenomenon in the Baradian sense, we can also understand it as effecting alternative practices that reconfigured boundaries between expert male physicians and docile female patients. Specifically, the intra-actions described earlier helped construct a new boundary, one that established a difference between women as subjects in need of—and capable of accessing—knowledge about their own bodies and women's bodies as knowable objects worthy of self-exploration. By changing what the patient *did*, *OBOS* changed who the patient *was*: she was a woman who possessed specialized knowledge of her own body. As such, *OBOS* demystified expert knowledge and destabilized normative boundaries, thereby exposing differences between doctors and patients—and the relations of power these differences authorized—as contingencies capable of being changed.

It is important to remember, however, that this subject-object split (in other words, woman-as-knower versus her body as object-of-knowing) was a contingent difference whose conditions of possibility emerged from specific material-discursive intra-actions. And like all cuts, this one excluded other possibilities. As Barad (2007) explains, "the enactment of boundaries . . . always entails constitutive exclusions and therefore requisite questions of accountability" (p. 135). So, while this particular cut helped women claim control over their own bodies and thereby reconfigure boundaries and create new habits of knowing, it excluded other kinds of cuts within the category of "woman" itself. Assuming an intersectional stance, Wells (2010) explains that early editions of *OBOS* posited women's bodies as "an undifferentiated material basis for feminism" (p. 91); in so doing, they failed to account for the ways in which women's access to and experiences with health care and medicine were shaped by race and class (pp. 91–93). Critiques of *OBOS* by women of color helped the collective realize that "knowing one's body is not enough" when it comes to reconfiguring boundaries in the politics of Western medicine (p. 95). Later versions attempted to account for prior exclusions by describing "women's experiences as they varied across classes, races, and ethnicities" (p. 94). By historicizing and examining both the motives and effects of this particular subject-object cut, Wells makes recognizable the ways in which the exclusions of *OBOS*'s became the conditions of possibility for its revision.[6]

Yet through a new materialist framework, we must also recognize that simple acts of inclusion fail to address how experiences of embodied difference materialize through specific intra-actions. Thus, the inclusion of perspectives by working-class women or women of color in *OBOS* won't create new habits that replace the systemic and enduring practices constitutive of discriminatory health care; rather, such change will emerge only through new kinds of material-discursive intra-actions, such as those that enjoin medical practitioners and hospital administrators in enactments that reconfigure boundaries between, to take one example, the insured and uninsured. Could *OBOS* be digitally remixed and used as a prop that changes the way triage nurses first encounter emergency room patients? We don't know. Feminist rhetorics are alive to possibilities of becoming.

A second study of activist rhetorics that helps demonstrate how scholarship in feminist rhetorics can contribute to work in feminist new materialism is Phaedra C. Pezzullo's (2007) *Toxic Tourism: Rhetorics of Pollution, Travel, and Environmental Justice*. Whereas our reading of Wells's book suggests how a feminist text understood as a Baradian phenomenon can produce agencies of change, we see Pezzullo's book as modeling how feminist scholar-researchers can participate in the production of such agencies. Toxic tours, Pezzullo explains, are appropriations of commercial tourism, "noncommercial expeditions into areas that are polluted by toxins, spaces that Robert D. Bullard calls 'human sacrifice zones'" (p. 5). The tours are organized and hosted by local community members who are motivated by their "collective desire to survive and to resist toxic pollution through active participation in public life" (p. 6). A toxic tour thus functions as a form of concrete political action that publicly renounces nonconscious habits of domination.

One such tour is set in San Francisco during October as a way to resignify what National Breast Cancer Awareness Month (NBCAM) might mean. Pezzullo explains that the

> San Francisco Bay Area of California has the highest rate of breast cancer of any area in a Western country. For women under forty, a predominately African American community in the Bay Area, Bayview / Hunters Point, has the highest breast cancer rate in the country. (p. 113)

Organized annually by the Toxic Links Coalition, the "Stop Cancer Where It Starts Tour" complicates dominant practices of NBCAM, which, while importantly raising awareness about breast cancer and supplying women with information about early detection, fail to ask, Why have we not "done more to stop the *sources* of environmentally linked cancers, particularly breast cancers"? (p. 111).

Through a feminist new materialist framework, we can conceptualize the intra-actions constitutive of NBCAM (pink ribbons, commonplaces such as "detection is your best protection," mammogram technologies, and so on) as making several cuts, one of which is a distinction between knower (doctors, women who conduct self-exams) and the object of knowing (early signs of breast cancer).[7] The intra-actions of the Toxic Links Coalition tour attempt to enact a different cut, one that selects the object of knowing as breast cancer's *causal sources*. Pezzullo describes some of these intra-actions from her location as a participant-observer on a 2001 tour, which made four scheduled stops at corporations responsible for operating and supporting toxic power plants. Between these stops, where local environmental justice activists delivered speeches, there were also creative performances and spontaneous encounters on the street with strangers, all of which gave rise to "an inventive, spontaneous, persuasive, and risky mobile theater" that Pezzullo argues is integral to the tour's embodied rhetoricity (pp. 115–116). One such encounter occurred before the tour began:

> I notice a shiny black sport utility vehicle pull up on the sidewalk.... The driver [named RavenLight] then walks in front of the police line, proceeds to unbutton her dress, pulls out her right arm, and exposes her mastectomy scar.... The police, of course, cannot stop or detain the woman in red for indecent exposure, because although it is illegal to bare a woman's breast in public, she has exposed no breast....
>
> ... A young woman steps in between RavenLight and myself. When she sees RavenLight's chest, she gasps. We stop. RavenLight glances back in the woman's direction. The young woman then reaches one hand out in the direction of RavenLight's exposed scar as she brings her other hand to her own chest, which is covered with a T-shirt that sinks to her touch.

Her eyes well up with tears and she says, "Sister, you are so brave." ... In that moment, all three of us—the woman in red who risks contact, the younger woman who risks reaching out to communicate, and the observer who risks sharing that intimate exchange—appear *present*. (pp. 119–120)

As a "mode of advocacy," Pezzullo's concept of presence "suggests that the materiality of a place promises the opportunity to shape perceptions, bodies, and lives" (p. 9); by feeling present, Pezzullo argues, the toxic tourist can begin to identify with and care about the lives and struggles of others. Through her experience of shared presence, she goes home to a different world, one that is connected to that other place of which she now feels a part. Pezzullo is careful to point out that because the toxic tourist can, after all, "go home," she enjoys privileges her hosts do not. Yet, for Pezzullo, even momentary contact helps "create an embodied rhetoric of identification that, in turn, motivates tourists to reimagine who and what matters to their community" (p. 139). In this way, toxic tours understood as Baradian phenomena help reconstitute habits of mattering.

Pezzullo's embodied rhetoric of identification also introduces the important role of presence in what Gilles Deleuze and Félix Guattari term "intercorporeal transformation." As Laurie Gries (2015) explains,

> intercorporeal transformation occurs when an entity experiences a transformation in identity, sense of purpose, position in society, or relations even as its body may not undergo significant change on the outside that can be detected. One commonly stated example is the transformation that a defendant being charged with murder experiences when a judge passes down a guilty verdict. (p. 63)

Although this definition suggests that transformation occurs on an individual level, Gries explains that the "transformative feeling" can become "infectious, leading to reassemblage of collective life" (p. 64). Returning to Pezzullo, we might say that the toxic tourist's embodied rhetoric of identification leaks out into the world: it sparks a reimagining of what matters in more bodies than one.

A feminist new materialist rhetorical framework extends Pezzullo's inquiry by asking, How are we to understand this motivation to reimagine who and what matters? One way to do so involves conceptualizing the tour as a phenomenon that reconfigures boundaries such that new agencies of change emerge. Doing so thus complicates any understanding of motivation that reduces it to a question of conscious human intention. As Barad (2007) explains, "cuts are agentially enacted not by willful individuals but by the larger material arrangement of which 'we' are a 'part'" (p. 178). On this view, human motives *become* via discursive-material entanglements and are thus irreducible to bodies, objects, or discourse. Furthermore, in the specific scene of Pezzullo's encounter, we can understand the *motivated* human as a distinction rendered intelligible by the intra-actions in which the *observing* human participates. By enacting practices that reconfigure what it means to "become aware," the toxic tour's intra-actions change who the toxic tourist is: she is a vulnerable body making contact and becoming someone else. As a toxic tourist herself, Pezzullo-the-researcher likewise enacts new practices of scholarly being. As her book becomes a rhetorical phenomenon, these practices leak out into the world, calling into being new collectives of feminist activist researchers who are in the process of becoming-thinking otherwise.

Taken together, Wells's and Pezzullo's projects emphasize the need to sustain vital connections between feminist research in the rhetorics of science and medicine. And, like Langford and Speight's study of #BlackLivesMatter, they also help demonstrate how projects that investigate activist efforts to mobilize political change have much to contribute to scholarship in feminist new materialism. Collectively, these projects suggest how feminist new materialist rhetorical practices can be understood as *inventive intra-actions*: making things, changing systems of relating, refiguring boundaries, enacting new practices of mattering, forging new habits. Such a theory of rhetorical invention is consistent with FRSS's posthumanist tenor, for it does not reposition the individual human actor as the sole origin of action. Instead, it conceptualizes possibilities for change as "emergent phenomen[a] that can only be directed, not determined" (Trader, 2013, p. 226). We participate in this directing by helping to enact some boundary-making practices to the exclusion of others. Importantly, then, intra-action retains the possibility of and the need for feminist intervention in the work of social justice. Indeed, through it inventing change emerges as both a political possibility and an ethical obligation. As Barad (2007) explains, "intra-acting responsibly as part of the world means taking account of the entangled phenomena that are intrinsic to the world's vitality and being responsive to the possibilities that might help us and it flourish" (p. 396). It is this position of openness, this "being alive to the possibilities of becoming" (p. 396)—and being accountable for and responsive to that which normalizing habits of mattering exclude—that constitutes the ground for feminist rhetorical action in a posthumanist world.

Notes

1. For thoughtful and compelling response to Washick and Wingrove's critique, see Bennett, 2015.

2. Indeed, as we write this, we learn of yet another police shooting of an unarmed African American man—Charles Kinsey, a behavioral health worker shot three times by Miami police while trying to help a patient. As NPR reports,

> Kinsey says he was lying on the ground with his hands up before he was shot—an account that's backed by the video that captured the moments prior to the shooting. . . . Kinsey later spoke to local TV news Channel 7 WSVN from his hospital bed to describe both his fears for his patient—who was apparently playing with a toy truck in the street—and his surprise at being shot.
>
> Kinsey told Channel 7: "I just got shot! And I'm saying, 'Sir, why did you shoot me?' and his words to me, he said, 'I don't know.'" (Chappell, 2016)

The "I don't know," we suggest, points to the officer's action as being both normalized and nonconscious. That the officer claimed to be aiming at the patient does not alter this interpretation, we would argue, but shows the same normalized and nonconscious response to people with disabilities. A recent report by the Ruderman Foundation found that "at least one third of all police killings are people with disabilities," suggesting a similarly serious situation for disabled people as for Black and Brown people (Disabled, 2016).

3. In an endnote, Barr also gestures toward the ways in which a nonhuman technology—a microphone—participated agentially in the emergence of credible scientific data during the Truvada hearings. Specifically, she explains that while the testimony of institutionally affiliated rhetors often exceeded three minutes, this was not the case for community rhetors, for whom the microphone was turned off after three minutes.

4. *OBOS* continues to be published and updated; however, Wells's book only addresses the editions created from 1973–1984.

5. In her overview of research in FSS, Banu Subramaniam (2009) situates feminist histories of science in a category she terms "Culture of Science" because these histories focus on "how the culture of science was made" (p. 961). One such history that models a rhetorical approach is Jordynn Jack's (2009) *Science on the Home Front: American Women Scientists in World War II*, which examines the making of scientific culture via analyses of its key commonplaces. Feminist rhetorical histories of science also consider the ways in which

nonscientists have historically appropriated the authority of scientific knowledge to build credibility and garner support for their arguments. One recent example is Wendy Hayden's (2013) *Evolutionary Rhetoric: Sex, Science, and Free Love in Nineteenth-Century Feminism*, which recovers feminist activists' appropriations of scientific theory in public argumentation and examines the broader social-cultural effects of those appropriations.

6. Wells also notes that the subject-object split sponsored by the collective's "rhetoric of second-wave feminism," which "could treat the body as an object owned by an autonomous subject who exercised rights over it," was in interesting ways

> at odds with the text's sense of the body and the subject as continuously linked in complex reafferent loops, in reflexive relations of causality and mutual implication. The collective wrote about a porous body, but they thought they were asserting the monologic sovereignty of the choosing of self over the body, seen as an object of possession. (pp. 172–173)

Wells carefully observes, however, that the second-wave feminist cut "express[es] the possibilities of a particular time and a particular social formation. Nothing that the collective wrote was intended to endure forever, or to work for all women everywhere" (p. 174).

7. Pezzullo's discussion of NBCAM, in terms of both its formation and effects, is substantive and complex. She considers, for example, the ongoing sponsorship of NBCAM by the billion-dollar pharmaceutical company AstraZeneca, which "'manufactures the world's best-selling cancer drug'" and was "at one point . . . the third-largest producer of pesticides in the United State" (p. 114). She also acknowledges that "one end result of NBCAM could be saving women's lives" and that "as NBCAM has grown exponentially, more people than ever before have begun to talk about breast cancer, a feminist accomplishment in itself" (p. 110).

References

Alcoff, L. M. (2006). *Visible identities: Race, gender, and the self.* Oxford, UK: Oxford University Press.

Barad, K. (2003). Posthumanist performativity: Toward an understanding of how matter comes to matter. *Signs, 28*(3), 801–831.

Barad, K. (2007). *Meeting the universe halfway: Quantum physics and the entanglement of matter and meaning.* Durham, NC: Duke University Press.

Bennett, J. (2015). Ontology, sensibility, and action. *Contemporary Political Theory, 14*(1), 82–89.

Boyle, C. (2016). Writing and rhetoric and/as posthuman practice. *College English, 78*(6), 532–554.

Chappell, B. (2016, July 22). North Miami office was aiming at man with autism, union chief says. *NPR.* Accessed 3 July 2017. Retrieved from http://www.npr.org/sections/thetwo-way/2016/07/22/487027848/north-miami-officer-was-aiming-at-man-with-autism-union-chief-says

Connolly, W. E. (2010). Materialities of experience. In D. Coole & S. Frost (Eds.), *New materialisms: Ontology, agency, and politics* (pp. 178–200). Durham, NC: Duke University Press.

Coole, D., & Frost, S. (2010). Introducing the new materialisms. In D. Coole & S. Frost (Eds.), *New materialisms: Ontology, agency, and politics* (pp. 1–43). Durham, NC: Duke University Press.

Disabled people account for 1 in 3 of all excessive force claims against police—study. (2016, March 16). *RT.* Accessed 3 July 2017. Retrieved from https://www.rt.com/usa/335872-police-brutality-disabled-population/

Engman, A., & Cranford, C. (2016). Habit and the body: Lessons for social theories of habit from the experiences of people with physical disabilities. *Sociological Theory, 34*(1), 27–44.

Gries, L. E. (2015). *Still life with rhetoric: A new materialist approach for visual rhetorics.* Boulder, CO: Utah State University Press.

Hansen, J. (2013). From hinge narrative to habit: Self-oriented narrative psychotherapy meets feminist phenomenological theories of embodiment. *Philosophy, Psychiatry, and Psychology, 20*(1), 69–73.

Hawk, B. (2011). Reassembling postprocess: Toward a posthuman theory of public rhetoric. In S. I. Dobrin, J. A. Rice, & M. Vastola (Eds.), *Beyond postprocess* (pp. 75–93). Boulder, CO: Utah State University Press.

Hayden, W. (2013). *Evolutionary rhetoric: Sex, science, and free love in nineteenth-century feminism.* Carbondale: Southern Illinois University Press.

Jack, J. (2009). *Science on the home front: American women scientists in World War II.* Urbana: University of Illinois Press.

Jagger, G. (2015). The new materialism and sexual difference. *Signs, 40*(2), 321–342.

Langford, C. L., & Speight, M. (2015). #BlackLivesMatter: Epistemic positioning, challenges, and possibilities. *Journal of Contemporary Rhetoric, 5*(3/4), 78–89.

Pezzullo, P. C. (2007). *Toxic tourism: Rhetorics of pollution, travel, and environmental justice.* Tuscaloosa: University of Alabama Press.

Royster, J. J., & Kirsch, G. E. (2012). *Feminist rhetorical practices: New horizons for rhetoric, composition, and literacy studies.* Carbondale: Southern Illinois University Press.

Subramaniam, B. (2009). Moored metamorphoses: A retrospective essay on feminist science studies. *Signs, 34*(4), 951–980.

Trader, K. S. (2013). Assuming differently: Posthumanism, enthymeme, and the possibility of change. *JAC, 33*(1/2), 201–231.

Washick, B., & Wingrove, E. (2015). Ontologized agency and political critique. *Contemporary Political Theory, 14*(1), 63–79.

Wells, S. (2010). *"Our bodies, ourselves" and the work of writing.* Stanford, CA: Stanford University Press.

Contributors

Index

Contributors

Liz Barr earned her PhD from the University of Wisconsin–Madison in the Rhetoric, Politics, and Culture Program of the Department of Communication Arts. Drawing on her training in gender and women's studies as well as rhetorical studies, Liz's research explores the interplays of public memory, queer politics, and scientific AIDS activism. Her work has appeared in *Feminist Collections*, the *Journal of Health Disparities Research and Practice*, and the *Journal of the International AIDS Society*.

Jennifer Bay is an associate professor of English at Purdue University, where she teaches courses in rhetorical theory, professional writing, feminist rhetorics, and community engagement. Her work has appeared in journals such as *College English, JAC, Programmatic Perspectives*, and *Writing Instructor*, as well as in edited collections.

Amanda K. Booher is an assistant professor of English at the University of Akron and specializes in the rhetorical, theoretical, and medical relationships of bodies and prosthetics. Her research has appeared in journals such as *Disability Studies Quarterly*, the *International Journal of Feminist Approaches to Bioethics*, and *Present Tense*.

Daniel J. Card is a PhD candidate in the English department at the University of Wisconsin–Milwaukee. There he studies rhetorics of science, technology, and medicine. His current work explores public participation in scientific and medical policy-making. His recent collaborative work was published in the *Annals of Internal Medicine* and *Qualitative Health Research*.

Catherine Gouge is an associate professor in the English department at West Virginia University. Her recent scholarship has appeared in the *Journal of Medical Humanities*, the *Journal of Technical Writing and Communication*, and *Rhetoric Society Quarterly*. Her research interests include the rhetoric of health and medicine, science and technology studies, writing and editing pedagogy, and technical communication.

S. Scott Graham is the director of the Public Engagement and Science Communication Laboratory and an associate professor at the University of Wisconsin–Milwaukee. He published *The Politics of Pain Medicine: A Rhetorical-Ontological Inquiry* in 2015 and is the author or coauthor of rhetorics of science and medicine scholarship in a variety of journals including the *Annals of Internal Medicine*, the *Journal of Medical Humanities*, and *Rhetoric Society Quarterly*.

Contributors

Jordynn Jack is a professor of English and comparative literature at the University of North Carolina at Chapel Hill, where she teaches rhetorical theory, rhetoric of science, health humanities, and women's rhetorics. She is a codirector of Health Humanities: An Interdisciplinary Venue for Exploration. Her books include *Autism and Gender: From Refrigerator Mothers to Computer Geeks* and *Science on the Home Front: American Women Scientists in World War II*. Her articles have appeared in journals including *College English*, *Quarterly Journal of Speech*, *Rhetoric Review*, and *Rhetoric Society Quarterly*.

Julie Jung is a professor of English at Illinois State University, where she teaches courses in contemporary rhetorical theories and writing, and the author of *Revisionary Rhetoric, Feminist Pedagogy, and Multigenre Texts*. Her scholarship has appeared in edited collections and journals such as *College English*, *Disability Studies Quarterly*, *Enculturation*, and *Rhetoric Review*.

Molly M. Kessler received her PhD from the English department at the University of Wisconsin–Milwaukee. She is currently an assistant professor at the University of Memphis. Her current work addresses the intersections of the rhetoric of health and medicine, chronic illness, patient experience, and medical technologies. She has published in *Body Image*, the *Journal of Health Communication*, and *Rhetoric Society Quarterly*.

Alex Layne is an assistant professor of technical communication at the Metropolitan State University in Minnesota. Her work focuses on gender, technology, and video games, and she created Minnesota's first Game Studies Program. She has published on harassment and advocacy in the video game industry as well as on procedural ethics, a methodology for studying meritocratic technological environments that forefronts ethics.

Jen Talbot is an assistant professor of writing in the School of Communication at the University of Central Arkansas. Her research interests include feminist materialist rhetorics, embodiment and affect, and institutional rhetorics. Her writing has appeared in *JAC* and *Works and Days*; she is currently focused on developing curriculum and building inclusive infrastructure for writing instruction.

Kyle P. Vealey is an assistant professor of English at West Chester University. His research and teaching interests include professional and technical communication, rhetoric of science, rhetorical theory and history, and civic rhetoric. His work has appeared in *Business and Professional Communication Quarterly*, the *Journal of Interactive Technology and Pedagogy*, the *Journal of Technical Writing and Communication*, *Programmatic Perspectives*, and *Rhetoric Review*.

Index

Italicized page numbers indicate figures and tables.

abortions, 93, 107–108
accountability: distributed, 107–108; maternal, 101, 103, 105–106; matters of, 64–69; as sociopolitical notion, 88; in treatment decisions, 110n13
Activity Theory (AT), 143, 145, 152–153
Actor-Network Theory (ANT): as critique of sociology's traditional methodologies, 85–87; Display Requirement and, 95–102; distributed accountability, 107; limitations of, 90–91; mappings, 102–103; as methodological approach in Technical Communication, 152; in understanding interconnected relations among technology, rhetoric, and human users, 142–143
ACT UP (AIDS Coalition to Unleash Power), 206
agency: accountability as inverse of, 88; defined, 86; distributed, 108; identifying locus of, 134n9; of matter, 145
agential realism: ANT and, 85–86; Barad's theory of, 126–127; Display Requirement and, 102–106; distributed work practices and, 155–157; and ethics of inclusivity, 108; personhood from perspective of, 107; phenomenon as basic ontological unit in, 89; qualitative interviewing study and, 145
Ahmed, S., 61, 64, 124–125
AIDS activism, 221
Alaimo, S., 133n4, 185
alien phenomenology, 74–75

anatomo-clinical method, 118–119
animacy concept, 133n3
animals, nonhuman, mattering of, 6
antiabortion rhetoric of Reagan era, 85
assemblagic agency concept, 228
associative theories, 152–153
Avastin hearings, 192–193

Barad, K.: account of agency as materially and discursively intra-active, 126–127; on agentially enacted cuts, 242; on Butler's theory of performativity, 42n15; on Butler's version of matter, 106; on co-constitution of the social and the scientific, 180; on constitution of reality by phenomena, 235; critique of the discursive, 102; emphasis on mutual constitution of agencies, 228; on ethics-matter, 68–69; on fetal enactments, 102–103; on inhuman versus nonhuman, 141–142; on inseparability of objects/subjects/time/ethics, 66–67; on materialization of all bodies, 184–185; on maternal accountability, 105–106; on phenomena's intra-activity, 66; on posthumanism, 30–31; on rhetorical work of experiments, 167–168; on theories and theorizing, 160; theory of entanglement, 169. *See also* agential realism
Barnett, S., 42n13, 51–52
becoming-thinking otherwise, rhetorically, 227

biological determinism critiques, role of, 21–22
biological essentialism, 185
biomedical activism, 205–208
biomedicine: autonomous subject presumption in, 125–126; calls for changing culture of, 135–136n15; compliance concept, 116, 118, 123, 133n3; concept of standardized body, 118–119, 123; discourse and ethos development, 211–212; predictive modeling, 125; scientific patriarchy in, 187
biopolitics, neoliberal, 121
Black, Maria (pseudonym), 216–217
#BlackLivesMatter, 229–230, 234–236
bodily autonomy, 84–85, 97–98
body, the: "ideal" and "normal" conceptions of, 122; Latour's definition and conceptualization of, 95–96; OOO and, 67–68; postmodernists and, 187; standardized model in Western biomedical culture, 118–119, 123; utilitarian concept as basis for compliance expectations, 124; in the world, 216–219
body-object-language model, 96
Bogost, I., 50–51, 58, 60, 75–76
book title, unpacking of, 3–4
Boston Women's Health Collective, 237–239
Boyle, C., 27, 229, 231
Braidotti, R., 21, 54, 59
Bryant, L. R., 50, 57, 60, 72–73, 184
Butler, J., 42n15, 61, 99, 102, 106

calibration, as term, 198–199
care in health and medicine, opportunities for, 127–128
Casper, M., 88, 99, 107
citation practices: APA guidelines, 8n1; Bryant on issue of, 72–73; critiques of, in feminist work in STS, 60–61; Haraway and, 68; importance in cultivating interdependence and dialogue, 71; and meaning, 63; politics in, in context of OOO, 61–62; reverberations into the past, 70–73; in scholarly work, 52–53
clinic ontology, 194–195
cognitive neuroscience, 167
compliance and noncompliance, rereading, 114–117
compliance concept, 116–118, 120–125, 128–129, 133n3
compliance metrics, communication and, 126
Compliance 1.0, 118–119
Compliance 2.0, 119–122
Condit, C., 26, 40n5, 94, 164, 168
constructivism, 21
constructivism, social, 1, 88–91, 133n4
Coole, D., 1, 183, 230
corporeal rhetoric, 209–210
correlationism, 56–57
coworking, 145, 152–155
cultural studies, 144
cuts or boundary-making practices, 38
cyborg figure, 7–8, 20, 23

deficit-model assumption, disability scholars and, 116
discourse and matter, rethinking relationship between, 1
discursive practices, defined, 31
disease/illness dichotomy, 186
diseases, enactments of, 187–188
Display of Real-Time View Requirement (Display Requirement): agential realism and, 102–106; ANT and, 95–102; distributed agency

and conceptual possibility of fetal personhood, 94–95; fetal ultrasound imaging and, 91, 232; posthumanist philosophies and, 90; site of decision-making in medical-legal apparatus, 98; unconstitutionality of, 84, 87, 91–92
distributed agency, defined, 109n2
distributed work practices, 151, 153, 155
Dolmage, J., 117, 122
Dred Scott decision (1857), 85
Drug Advisory Committees (DACs), FDA: antiviral, 210–212; evaluation of meetings (2009–2012), 190–192; FDA and, 189–190; individual stakeholder contributions per meeting, *191t*; inter-rater reliability, *196t*; medical experts' role in, 189–191; member type distribution, *191t*; oncologic, 192–193; ontologies by utterance, most frequently occurring, *196t*; program revisions, 199–200; PR program, 188–190, 198–199

ecological thought, 57–58
embodied beings, 67
embodied rhetoric, 209–210, 241
embodied vernacularity: community rhetors and, 206–207, 216–217; focus of, 209, 218–219; as form of situated knowledge, 208; institutional voices and, 210–212; theory of, 39; in Truvada hearings, 217–218, 220–221
empathic accuracy, gender differences and, 166, 170–172
Enlightenment legacy, 64
entanglement theory, 169
epistemic neglect, 7
equality rhetoric, 115

ethical intervention, as aim of rhetorical-cultural study, 219
ethical practice, approach to ontologies and, 77–78
ethics: in acknowledgments and citations, 72; entanglement with ontology and epistemology, 53; of inclusivity, agential realism and, 108; inseparability of objects/subjects/time and, 66–67; obligation of scholars, 6; in ontology of the world, 67; in response to divergent behaviors by health-care participants, 132; of responsiveness, 210
evidence-based medicine, 193–194
experiments: effect of material elements of, 170, 172, 177; feminist rhetoricians and, 179; gender-in-the-making within, 169, 173–174, 178–179; rhetorical work of, 167–168

Fahnestock, J., 4, 9n5, 164
Fausto-Sterling, A., 164–166, 176
FDA (U.S. Food and Drug Administration): Antiviral Drugs Advisory Committee, 210–212; failure to recognize embodied expertise of patients, 192; hearing on Gilead Sciences' new drug use application for Truvada, 206; Oncologic Drugs Advisory Committee, 192–193. *See also* Drug Advisory Committees (DACs), FDA
feminism, 50–54, 86, 205–208
feminist embodiment, 208
feminist materialisms, ANT and, 102–103
feminist materialist cultural theory (new materialists), 22
feminist methodology of engagement, 77

feminist new materialism, 20–23, 154, 159, 183–186, 227–229, 236–242
feminist posthumanism, 1, 6, 27–32, 88–89. *See also* feminist new materialism
feminist rhetorical science studies (FRSS): adoption of NM in, 183; citation practices issue, 52–53; embodied vernacularity and research in, 207; goals of, 1; need for, 141; ontological status of objects, 65–66; as response to call for hybrid approaches, 24
feminist rhetorical theory, linkage with FSS, 3–4
feminist rhetorics: habit conceptualization, 228–231; materiality in intersections of rhetorics of science and, 1–2; methodology, 53–54, 69; of personal autonomy, 87–88; practices of, as doings otherwise, 229; scholars of, 53, 239–240
feminist scholars and scholarship: archival research, 2; in brain sciences, 164–166, 178, 180; critiques of science, 19; exclusion from OOO's philosophical history, 52, 55, 63, 78n2; research in rhetorics of science and medicine, 179, 242; Richardson on, 8n4; in technical communication journals, 147–148
feminist science studies (FSS): alliance between rhetorical inquiry and, 5; attention to material concerns, 20; emerging scholarship in, 1; focus on text-based critiques of established biological research, 24–25; influences on, 9n5; materiality and relationality of bodies involved in knowledge production, 207; overview, 18–20; posthuman view, 67; purpose of introduction to, 40n1; questions of boundedness and meaning, 68; tackling of barrier to interdisciplinarity between humanities and natural/physical sciences, 3–4
feminist technoscience, in FSS, 20
feminist theory impasse, 133n4
fetal agency, posthumanist theories and, 85–86
fetal life support, pregnant woman as, 100
fetal personhood: and accountability for infrastructure of care, 106; efforts to establish legal standards for, 88, 91–95, 100–102, 104, 110n12; pregnant woman's role as mediator of, 94; rejection of, prior to fetal ultrasound technologies, 109n1
fetal ultrasound imaging: as form of biotourism, 100; mediating force of, 98–99; narration by technicians, 93–94; role in construction of fetus as subject, 84–86, 91–92, 100–101; two-dimensional versus three-dimensional, 93. *See also* Display of Real-Time View Requirement (Display Requirement); fetal personhood
Fine, C., 164–165
flat ontologies, 57, 84–87, 95
flesh, theory of the, 208, 215
fMRI (functional magnetic resonance imaging), 167
Foucault, M., 42–43n16, 118–119
Frost, S., 1, 21, 183–184, 230

gender: academic inequities of race, class, and, 144; dynamics of, in male-dominated industries, 149; keyword searches on, 147; and objects, 152–158; and status of

technical writers employed (2010), *148f*; of technical writers by education levels (2010), *149f*; undergraduate students and complexities of, 146–147. *See also* sex/gender differences
gender-aware methodologies, 168–173
gendered attitudes, cognitive processes and, 176
gendered experiences, in Technical Communication, 143, 145–151
gender-linked concepts, 171–172
genealogical methodology, 42–43n16
genomic research, 120–121
genre-discourse analysis, 25
Gilead Sciences, 206, 222n3
Graham, S. S., 126, 192–193, 199
Gries, L. E., 23, 235, 241
Grosz, E., 22, 59–60

habits, 228–231, 234
habitual practices, 231–235, 237–239
Halberstam, J., 54, 58–59
Happe, K. E., 120–121
Haraway, D.: on cat's cradle as trope, 71; citations by, 68; and cyborg figure, 7–8, 20, 23; on exclusion of women from scientific work, 63; on rhetorical vision, 217; situated knowledges, 165, 208; view from somewhere, 215
Harding, S., 19, 165
Harman, G., 50, 56–57, 59–60, 64–65, 75
hashtag, as rhetorical actant, 234–235, 236
#BlackLivesMatter, rhetorical analysis of, 229–230, 234–236
Hawhee, D., 135n14, 217–218
Hayles, N. K., 20–21, 67–68
health and seronegativity, conflation of, 222n2

health-care participants. *See* patients
health-care professionals, practices of, 126
health policy decision-making, post-plural approach to, 198–199
Heidegger, M., 50, 75
Hekman, S., 133n4, 145, 154, 185
heuristics, as material framework for meaning, 160–161
HIV, 205, 214, 219
Hodges, S. D., 166, 170–171
home ontology, 194, 197–198
hybrid rhetorical strategies, 213
hyperobjects, emergence of, 58

individual characteristics, erasure through scientific/institutional discourse, 211
informed consent, 92
intercorporeal transformation, 241
intersectionality, 8–9n4
intra-actions, 30, 31–32, 66–67, 130, 159–160, 230, 242

Jack, J., 36–37, 233, 243–244n5
Johnson, J., 220–221

kairologies of care, 116, 129–130, 136n16, 232
kairos, 131, 135n14
Keränen, L., 9n5, 23–24
Kirsch, G. E., 53, 69, 228–229
Klein, K. J. K., 166, 170–171

lab ontology, 193–194, 196–198
laboratory, material setting of, 143–144
Langford, C. L., 94, 229–230, 234
Latour, B.: ANT, 85–86; concept of the body, 95–96; exploration of reciprocal entanglements, 185; humanness as equal and

Latour, B. (*continued*)
  interchangeable, 144; models proposed in "Irreductions," 51; on postmodernism, 183–184; on role of citations, 62; "the social," as term, 87; on social asymmetries, 89–91; thing, etymology of, 75. *See also* Actor-Network Theory (ANT)
logic of care, 130–131, 136n17

mangle concept, 154–156
marginalized rhetors, historiographic approach to study of, 221
market ontology, 195–196
Martins, D. S., 118, 132n1, 134n9
material-discursive concept, 30–31
material-discursive entanglements, 126, 166. *See also* intra-actions
material feminists, and social constructivism, 88–91
materiality: feminist theory and, 185; feminist thought engagement with, 54; in intersections of feminist rhetorics and rhetorics of science, 1–2; posited by social constructivist theories, 133n4; symbolicity contrasted with, 23–24
material rhetorics, major strands in, 28
matter, postclassical theories of, 22–23
McConnell, Michael (pseudonym), 213–215
mechanical objectivity notion, in medical perception, 119
mediators, defined, 96
medical experts (MEs), role in DACs, 189–191
medicalization of bodies at risk, consequences of, 121
medical mystification, habitual practices of, 237–239
medical perception, 119
medicine, evidence-based, 118
methodologies: of apolitical description, 41n7; gender-aware, 168–173; genealogical, 42–43n16; of the oppressed, 7–8; of rhetorical reverberations, 5, 69–77
Mol, A., 62, 70, 96–98, 130–131, 136n17, 186–187, 198–199
Morton, T., 50, 57–58
motherhood, from agential realism perspective, 105
Murray, S. J., 121, 134n8

National Breast Cancer Awareness Month (NBCAM), 240–241
network, defined, 96, 98
neuroimaging research, sex/gender differences and, 164, 173–178
neurorhetorics, 166–168
neuroscience, 164, 167–168, 172, 179
new materialisms (NM): approach to representation in pharmaceutical policy, 199–200; associative theories distinguished from, 152; and feminist political projects, 185–186, 188; meaning of term, 183–184; ontology of becoming and, 231; postclassical theories of matter and, 22–23; radical feminist materialism of figures and, 59; scholars and focus of, 55; tenets of, 36
noncompliance: as deviant subject position, 122–123; as evidence of coping, 35, 131; problems with, 117–118; revaluing, 124–129; understanding, among health-care professionals, 114. *See also* patients
nonhuman entities, 85, 232, 243n2
Nordstrom, S. N., 145, 155–156, 158
normalcy rule, 122–124

null hypothesis significance testing, 164, 168
NVivo for Teams, 190

object-interview, as ontological event, 155–156, 158
object-oriented ontology (OOO): criticisms of, 52, 54, 58–65, 67–68, 78n2; as emerging movement, 50–54; interdisciplinary engagement with rhetoric studies, push for, 51–52; object, use of term, 75; ontological status of objects, 65–66; scholarship in, and purpose of approach, 2; tenets of, 56–58; thingness and, 55
object-oriented rhetoric philosophies, 1–2, 42n13
objects: agential relationships to one another, 50; of genealogy, 155–156; as phenomena, 235–236; as term, 75–77
object-subject, 158
Oncologic Drugs Advisory Committee, 192–193
onticology, 57
ontologies: approach to, as ongoing ethical practice, 77–78; enactment of, through practice, 69; flat, 51; present in Avastin public comment period, 192; scholarly practices and enactment of, 62
Our Bodies, Ourselves (OBOS) (Boston Women's Health Collective), 237–239
overmining, defined, 56–57

pain studies, gender roles and, 166, 169, 171–172, 199
Parikka, J., 54, 59–60
patient autonomy, bioethicists and, 134n8

patient-centered care movement, 116
patient discharge communication, 117, 126
patient representation, efforts of postmodern inquiry into, 186–187
patient representative (PR) program, FDA, 189–190, 198–199
patients: accountability in treatment of, 110n13; deviant, bodies as cautionary tales, 121–122; divergent behaviors by, 127–128, 130–132, 133n3; dynamic constitution of agencies and phenomena in, 127–129; FDA's failure to recognize embodied expertise of, 192; revaluing noncompliance by, 124–129
perceptions: habitual, 230, 234; medical, 119
personhood, 84–85, 94, 107. *See also* fetal personhood
Pezzullo, P. C., 239–242, 244n7
phenomena: choices and actions of health-care participants as consequence of convergence of, 127–129; emergent, 242; intra-active, convergences of, 66, 130, 157; objects as, 235–236; as primary ontological units, 89, 135n13
place, embodied and experiential aspects of, 154–155
politics of who versus politics of what, 2, 186–188, 189–192, 192–198, 198–200
posthumanism: absence of challenges to Western-centrism in, 6; attention to materiality, 87; central tenets of, 86; contemporary approaches, 21; and critiques of the social, 86–91; emergence of, in FSS, 20; feminist, 1, 27–32, 67, 88–89 (*see also* feminist new materialism); history of, 27; practice theory, 229;

posthumanism (*continued*)
  reproductive rights, feminism, and, 88; rhetorical practice understanding, 231; theories and fetal agency, 85–86
posthumanist rhetoric scholarship, 5
posthuman/object-oriented scholars, 2–3
postmodernism, 184, 186–187
Potter, Jacob (pseudonym), 218–219
power asymmetries, between scientific knowledge and community expertise, 212–215
practice coding schema, sites of, *193t*
praxiographic approach to disease, 188
pregnant bodies, public policing of, 103
PrEP (pre-exposure prophylaxis), 205–206, 214, 218–219, 222n1
presence concept, 241
professional writing, at Purdue University, 145–146
public rhetoric of science, focus of, 25

qualitative research, 155–156, 190

randomized controlled trial (evidence-based medicine), 193–194
RavenLight, 240–241
reading *Feminist Rhetorical Science Studies*, 5–8
Reagan era, antiabortion rhetoric of, 85
realist thought, OOO as emerging school of, 54
reflective practices, 73–77
relational ontology, 21
remission society, in modern medicine, 119–120
representationalism, 178, 184–185

reproductive rights, feminism, posthumanism, and, 88
research practices, unquestioned, 233
responsiveness, defined, 210
reverberations, 5, 54, 69–77, 70–75, 77
rhetorical actancy, 235
rhetorical analyses of scientific texts, limitations of, 26
rhetorical formation theory, 26
rhetorical invention theory, 242
rhetorical reverberations, 5, 54, 69–77, 70–75, 77
rhetorical theory, intersections of OOO and, 51–52
rhetorical vision, 217–218
Rhetoric of Science, Technology, and Medicine (RSTM), 3–5, 23–26
rhetorics: focus on, as dynamic phenomenon, 28; Indigenous philosophies and, 5–6; of science, 23–27
Richardson, S. S., 8–9n4, 19
Robinson, M. E., 166, 169
Roe v. Wade (1973), 85, 87, 94, 108, 110n12
Royster, J. J., 53, 69, 228–229

scholarship, social circulation of, 53
science, rhetorical critiques of, 24–25
science and technology studies (STS), 41n7, 183
scientific inquiry, as neutral arbiter of fact, 76
scientific knowledge production, 25, 165
scientific objectivity, 64
scoping and scaling, 74
Scott, J. B., 20, 24–27, 210, 219
sedimentation concept, 230, 234
sex/gender differences: assumptions in research on, 168; empathic accuracy study, 166, 170–172; human

motion study, 177–178; materialization of, 169, 178–179; neuroimaging research and, 164, 173–178; pain studies and, 199; scientific studies of, 164–166
sick role, in medical studies, 119, 134–135n10
situated knowledge, 68, 208
social constructivism, 1, 88–91, 133n4
social inequality, blackboxing questions of, 41n7
somatic mind theory, 30
speculative realism (SR), 56, 59
Speight, M., 229–230, 234
Spinuzzi, C., 145, 151–155
standpoint epistemology, 165
STEM, 141–142, 146, 158–159
Stern-Gerlach experiment, 157–158
strategic contemplation, defined, 69
strong objectivity theory, 19
subject-object binary, 141–145, 238–239
Subramaniam, B., 19, 24–25, 243–44n5

technical communication/Technical Communication: ANT as prominent methodological approach in, 152; approaches to distributed work in, 151; contributions of women in, 150–151; feminist new materialism and re-seeing intra-actions of human beings in, 159–160; feminist scholarship in journals of, 147–148; human materiality and, 141–145; impact of studying the nonhuman, 144–145; lack of attention to gendered experiences of practitioners, 146–147; methodology as technological apparatus, 157; on new materialist theory and women students in, 232–233; technical writers by gender and status (2010), 148f, 149f
Technical Communication Theory, 142
technologies, shifting understandings of term, 150
technoscience, defined, 20
teleology-of-care rationale, assumptions of, 124
Teston, C., 152, 164, 192–193
thing, as term, 75
thing theory, 54–55
Toxic Links Coalition, 240
toxic tours and tourists, 239–241
transnational technoscience scholarship, 27
Truvada hearings: Antiviral Drugs Advisory Committee composition, 210–212; approval for PrEP, 220; community testimonies, 207, 212; content of community rhetors' testimonies, 216–217; dominance of institutional voices in, 211–212; embodied vernacularity versus clinical trial data, 217–218, 220–221; enforcement of three-minute time limit for community testimony compared to scientific speakers, 213, 223n8, 243n3; focus of scientific testimony, 216; rhetorical strategies employed by vernacular actors, 206, 219

ultrasound technologies. *See* fetal ultrasound imaging
undermining, defined, 56
U.S. Census Bureau, trends in employment among technical writers, 148
U.S. Food and Drug Administration. *See* FDA (U.S. Food and Drug Administration)

vernacular discourse, embodied rhetoric and, 209–210
vernacularity, as hybrid position, 215
vernacular rhetors, strategies used by, 207
voice study, 177–178

Washick, B., 22–23, 228–229
Wells, S., 237–239
Wingrove, E., 22–23, 228–229
women: infantalization of, in academia, 63; as initiators of technomedical actions, 99; and Internet searches for "object," 76–77; lack of engagement in STEM fields, 141–142, 146, 158–159; pregnant, physiological and biochemical changes in, 104–105
Women's Health Collective (Boston), 237–239
women's movement (1970s), 19
Women's Right to Know Act (2011), 91
workplace, shifting understandings of term, 150

Studies in Rhetorics and Feminisms

Studies in Rhetorics and Feminisms seeks to address the interdisciplinarity that rhetorics and feminisms represent. Rhetorical and feminist scholars want to connect rhetorical inquiry with contemporary academic and social concerns, exploring rhetoric's relevance to current issues of opportunity and diversity. This interdisciplinarity has already begun to transform the rhetorical tradition as we have known it (upper-class, agonistic, public, and male) into regendered, inclusionary rhetorics (democratic, dialogic, collaborative, cultural, and private). Our intellectual advancements depend on such ongoing transformation.

Rhetoric, whether ancient, contemporary, or futuristic, always inscribes the relation of language and power at a particular moment, indicating who may speak, who may listen, and what can be said. The only way we can displace the traditional rhetoric of masculine-only, public performance is to replace it with rhetorics that are recognized as being better suited to our present needs. We must understand more fully the rhetorics of the non-Western tradition, of women, of a variety of cultural and ethnic groups. Therefore, Studies in Rhetorics and Feminisms espouses a theoretical position of openness and expansion, a place for rhetorics to grow and thrive in a symbiotic relationship with all that feminisms have to offer, particularly when these two fields intersect with philosophical, sociological, religious, psychological, pedagogical, and literary issues.

The series seeks scholarly works that both examine and extend rhetoric, works that span the sexes, disciplines, cultures, ethnicities, and sociocultural practices as they intersect with the rhetorical tradition. After all, the recent resurgence of rhetorical studies has been not so much a discovery of new rhetorics as a recognition of existing rhetorical activities and practices, of our newfound ability and willingness to listen to previously untold stories.

The series editors seek both high-quality traditional and cutting-edge scholarly work that extends the significant relationship between rhetoric and feminism within various genres, cultural contexts, historical periods, methodologies, theoretical positions, and methods of delivery (e.g., film and hypertext to elocution and preaching).

Queries and submissions:
Professor Cheryl Glenn, Editor
E-mail: cjg6@psu.edu
Professor Shirley Wilson Logan, Editor
E-mail: slogan@umd.edu

Studies in Rhetorics and Feminisms
Department of English
142 South Burrowes Bldg.
Penn State University
University Park, PA 16802-6200

Other Books in the Studies in Rhetorics and Feminisms Series

*Retroactivism in the
Lesbian Archives:
Composing Pasts and Futures*
Jean Bessette

*A Feminist Legacy:
The Rhetoric and Pedagogy
of Gertrude Buck*
Suzanne Bordelon

*Regendering Delivery:
The Fifth Canon and
Antebellum Women Rhetors*
Lindal Buchanan

Rhetorics of Motherhood
Lindal Buchanan

*Conversational Rhetoric:
The Rise and Fall of a Women's
Tradition, 1600–1900*
Jane Donawerth

Feminism beyond Modernism
Elizabeth A. Flynn

*Women and Rhetoric
between the Wars*
Edited by Ann George,
M. Elizabeth Weiser, and
Janet Zepernick

*Educating the New Southern Woman:
Speech, Writing, and Race at the
Public Women's Colleges, 1884–1945*
David Gold and Catherine L. Hobbs

Food, Feminisms, Rhetorics
Edited by Melissa A. Goldthwaite

*Women's Irony:
Rewriting Feminist
Rhetorical Histories*
Tarez Samra Graban

*Claiming the Bicycle:
Women, Rhetoric, and Technology
in Nineteenth-Century America*
Sarah Hallenbeck

*The Rhetoric of Rebel Women:
Civil War Diaries and
Confederate Persuasion*
Kimberly Harrison

*Evolutionary Rhetoric:
Sex, Science, and Free Love in
Nineteenth-Century Feminism*
Wendy Hayden

*Liberating Voices:
Writing at the Bryn Mawr Summer
School for Women Workers*
Karyn L. Hollis

Gender and Rhetorical Space in American Life, 1866–1910
Nan Johnson

Antebellum American Women's Poetry: A Rhetoric of Sentiment
Wendy Dasler Johnson

Appropriate[ing] Dress: Women's Rhetorical Style in Nineteenth-Century America
Carol Mattingly

The Gendered Pulpit: Preaching in American Protestant Spaces
Roxanne Mountford

Writing Childbirth: Women's Rhetorical Agency in Labor and Online
Kim Hensley Owens

Rhetorical Listening: Identification, Gender, Whiteness
Krista Ratcliffe

Feminist Rhetorical Practices: New Horizons for Rhetoric, Composition, and Literacy Studies
Jacqueline J. Royster
and Gesa E. Kirsch

Rethinking Ethos: A Feminist Ecological Approach to Rhetoric
Edited by Kathleen J. Ryan, Nancy Myers, and Rebecca Jones

Vote and Voice: Women's Organizations and Political Literacy, 1915–1930
Wendy B. Sharer

Women Physicians and Professional Ethos in Nineteenth-Century America
Carolyn Skinner

Praising Girls: The Rhetoric of Young Women, 1895–1930
Henrietta Rix Wood